Introduction to the
MATHEMATICAL THEORY
OF CONTROL PROCESSES

VOLUME II
Nonlinear Processes

This is Volume 40-II in
MATHEMATICS IN SCIENCE AND ENGINEERING
A series of monographs and textbooks
Edited by RICHARD BELLMAN, *University of Southern California*

A complete list of the books in this series appears at the end of this volume.

Introduction to the

MATHEMATICAL THEORY

OF CONTROL PROCESSES

Richard Bellman

DEPARTMENTS OF MATHEMATICS
AND ELECTRICAL ENGINEERING
UNIVERSITY OF SOUTHERN CALIFORNIA
LOS ANGELES, CALIFORNIA

VOLUME II
Nonlinear Processes

1971

ACADEMIC PRESS   New York and London

ACADEMIC PRESS, INC.
111 Fifth Avenue, New York, New York 10003

United Kingdom Edition published by
ACADEMIC PRESS, INC. (LONDON) LTD.
Berkeley Square House, London W1X 6BA

LIBRARY OF CONGRESS CATALOG CARD NUMBER: 67–23153

PRINTED IN THE UNITED STATES OF AMERICA

To Stan Ulam, friend and mentor

CONTENTS

Chapter 3 **Computational Aspects of Dynamic Programming**

Chapter 4   **Continuous Control Processes
and the Calculus of Variations**

Chapter 5   Computational Aspects of the Calculus of Variations

Chapter 6   Continuous Control Processes and Dynamic Programming

Chapter 7 Limiting Behavior of Discrete Processes

Chapter 10   **Abstract Control Processes and Routing**

Chapter 11   **Reduction of Dimensionality**

Chapter 12   **Distributed Control Processes and the Calculus of Variations**

Chapter 13 Distributed Control Processes and Dynamic Programming

Chapter 14 Some Directions of Research

PREFACE

In the first volume of this series, devoted to an exposition of some of the basic ideas of modern control theory, we considered processes governed by linear equations and quadratic criteria and discussed analytic and computational questions associated with both continuous and discrete versions. Thus typical problems are those of minimizing the quadratic functional

$$J(x, y) = \int_0^T [(x, Ax) + (y, By)]\, dt$$

where the vectors x and y are connected by the differential equations $x' = Cx + Dy$, $y(0) = c$, and its discrete counterpart. In this volume we wish to broaden the scope of these investigations.

First we shall consider the more general problem of minimizing a functional

$$\int_0^T g(x, y)\, dt$$

where x and y are related by the equation $x' = h(x, y)$, $x(0) = c$ and its companion discrete version, the question of minimizing the expression

$$\sum_{n=0}^N g(x_n, y_n),$$

where

$$x_{n+1} = h(x_n, y_n), \qquad x_0 = c.$$

This latter can be readily treated by means of the theory of dynamic programming.

We shall pursue a parallel development. Chapters 2 and 3 are devoted to analytic and computational aspects of the foregoing using dynamic programming. Chapters 4 and 5 are devoted to the same types of questions for the continuous process using the calculus of variations.

The order of presentation is the reverse of the order in Volume I where the calculus of variations was treated first. There the motivation was that a simple rigorous account could easily be given and that this approach could in turn be used to provide a rigorous derivation of the fundamental Riccati equation of dynamic programming.

Here we use dynamic programming to provide a simple rigorous approach to general discrete control processes. At the cost of essentially no additional effort, we can in this way handle constraints and stochastic effects. On the other hand, the consideration of these important aspects of general continuous processes requires a non-negligible mathematical training and sophistication.

The reader who has mastered the material in Chapters 2 and 3 is well prepared to consider a number of significant control processes that arise in biology, medicine, economics, engineering, and the newly emerging environmental sciences. We have avoided any discussion of applications here, although it is clear that our mathematical models and our continued emphasis upon numerical solution are both motivated by the kinds of questions that arise in numerous scientific investigations.

In Chapters 6, 7, and 8 we consider various types of interconnections, between different types of control processes, between discrete and continuous processes, between finite and infinite processes, and between dynamic programming and the calculus of variations. The application of dynamic programming to continuous processes introduces the basic nonlinear partial differential equation

$$f_T = \min_v \left[g(c, v) + v f_c \right],$$

an object of great interest in its own right. We have, however, not pursued the study in any depth since this probably involves the theory of partial differential equations much more than control theory. The reader interested in derivation of a number of fundamental results in the calculus of variations, plus Hamilton–Jacobi theory, directly from this equation may consult the cited book by S. Dreyfus.

In Chapters 9, 10, and 11 we turn to methods specifically aimed at the numerical solution of control problems: duality, reduction of dimensionality, and routing. In Chapters 12 and 13 we present a brief introduction to the study of distributed control processes, specifically processes governed by partial differential and differential-difference equations. Finally in Chapter 14 we very briefly indicate some of the vast, relatively unexplored fields of control theory awaiting study.

Throughout our aim has been to provide a relatively simple introduction to one of the most important new mathematical fields, the theory of control processes. Numerous references are given to books and research papers where more intensive and extensive study of dynamic programming, the calculus of variations, computational techniques, and other methods may be found. We have tried to preserve the impartial attitude of a guide through a museum pointing out the many fascinating treasures that it contains without dwelling overly on individual favorites.

I would like to express my appreciation to Art Lew who diligently read the manuscript and made many valuable suggestions, as well as to Edward Angel, David Collins, John Casti, Tom Higgins, and Daniel Tvey, and to Rebecca Karush who patiently typed a number of revisions.

1

BASIC CONCEPTS OF CONTROL THEORY

1.1. Introduction

The feasible operation and the effective control of large and complex systems constitute two of the central themes of our tumultuous times. Although considerable time, human and computer resources, and intellectual effort have already been devoted to their study, one can foresee that their dominating position in contemporary application and research will be maintained and even accentuated. As always, a nexus of problems of social, economic and scientific import creates its associated host of novel, intriguing and formidable mathematical problems. A number of these questions may be profitably contemplated and treated with the aid of the existing mathematical theory of control processes. Many, however, require further elaborations and extensions of this theory; many more appear to require new theoretical developments.

In this chapter we wish to lay the groundwork for much of the discussion and analysis that follows and to present some of the conceptual foundations for both discrete and continuous control processes of deterministic type. Both of these processes, needless to say, represent drastically simplified versions of realistic and significant control processes, which we shall discuss in some detail in Chapter 14. Nonetheless, they serve the most useful purpose of providing convenient starting points for analytic and computational discussions of didactic nature. They can readily be used to illustrate the advantages and disadvantages of both the calculus of variations and dynamic programming. Furthermore, a careful investigation of their numerous shortcomings provides us with points of departure for extensive research into the formulation and the treatment of more meaningful control processes.

We shall, in this and other endeavours, occasionally retread ground already covered in the first volume in order to make the presentation here as self-contained as possible. We feel that no serious harm is done by a certain amount of repetition.

1.2. Systems and State Variables

We shall employ the useful word "system" in a precise fashion to describe a combination of a state vector, $x(t)$, and a rule for determining the behavior of this vector over time. We shall be principally concerned in what follows with finite-dimensional vectors, although in later chapters we shall briefly consider infinite-dimensional vectors in connection with distributed control processes.

The two simplest, and therefore most common presumptions for future behavior, are expressed analytically either in terms of a difference equation,

$$x_{n+1} = g(x_n), \qquad x_0 = c, \tag{1.2.1}$$

or in terms of a differential equation,

$$\frac{dx}{dt} = g(x), \qquad x(0) = c. \tag{1.2.2}$$

In the first case, the variable we may consider to represent time is discrete-valued, $t = 0, 1, 2, \ldots$, in the second case, time is continuous.

In both formulations we make the basic assumption that the future depends only on the present and not at all on the past. This is more a question of the choice and dimension of the state vector than any intrinsic property of the system.

1.3. Discussion

What are the connections between this mathematical definition of the term "system" and the complex physical entities of the real world that we blithely call "systems"? There is no easy answer. One preliminary answer, however, is that a physical system can be, and should be, treated mathematically in many different ways. Hence, to any "real" system corresponds innumerable mathematical abstractions, or "systems" in the foregoing sense. We can consider the real system to include the set of all such realizations.

This one-to-many correspondence, this *cardinal* concept, cannot be over-emphasized. Too often, the student, who is familiar only with one conventional mathematical realization (a heritage of bygone eras), consciously and unconsciously identifies a particular set of equations with the real system. He may then promptly forget about the physical system and focus solely upon the mathematical system.

This narrowness of vision results in a number of losses. First, there is a loss of flexibility and versatility. Secondly, fundamental questions are easily overlooked in favor of mathematical idiosyncrasies. Thirdly, modern systems contain many basic features not present in classical formulations.

The mathematical realization utilized should depend critically upon both the type of results desired and the tools available for the production of these results. Many new problems arise in connection with the choice of an appropriate mathematical representation.

1.4. Control Variables

Descriptive equations of the type shown in Section 1.2 enable us to predict the future behavior of the system given a knowledge of its current state. On the basis of calculations of this nature, and occasionally even simply on the basis of observation, we can readily conclude that many systems do not perform in a completely acceptable fashion. There can be many reasons for our dissatisfaction: components of the system may begin to malfunction; the inputs to the system, as determined by the environment, may change in quantity and quality over time; and, finally, there may be significant changes in our own goals and evaluation of performance. All of these phenomena are particularly evident in the case of large systems, especially social systems.

A first thought in the search for a system exhibiting better performance over time is to introduce an entirely new system, or at least to rebuild the old system. The first alternative is generally not practical. For example, in the biomedical sphere it is not feasible to think of using an entirely different body run by the old brain, which is what one would consider to be the real identity. It is feasible, however, to use organ transplants, artificial kidneys, hearts and so forth, to alleviate certain medical conditions. The field of prosthetics and orthotics is one of the challenging new areas of control theory.

In the economic and social spheres, it is not feasible to contemplate the use of entirely different systems because we don't know enough to accomplish this change effectively without the introduction of transient effects, which can be both unexpected and quite unpleasant. History makes us all gradualists. If we don't understand very much about the operation of a complex system, changes must be introduced slowly and carefully.

Let us then turn our attention to a different approach, the concept of the use of additional and compensating effects which modify or remove undesired phenomena. We shall avoid in what follows the exceedingly complex engineering questions connected with precise ways in which these effects are actually introduced into the system. Clearly, these are basic problems of the utmost significance as far as the construction of meaningful mathematical models is concerned. Often, consideration of these questions leads to new and important mathematical problems. Furthermore, it must be admitted that sufficient attention has not been paid to these questions in the past. Some discussion of operational considerations will be found in Chapter 14.

Since, however, we are primarily interested in presenting an introductory treatment of the many mathematical areas of modern control theory, we will suppose simply that the net result of the various control activities in and around the physical system is the appearance of another function in the defining equations of Section 1.2.

Thus, in the discrete case, the descriptive equation takes the form

$$x_{n+1} = g(x_n, y_n), \qquad x_0 = c, \tag{1.4.1}$$

while the in continuous case we have

$$x' = g(x, y), \qquad x(0) = c. \tag{1.4.2}$$

We call this new function y appearing on the right-hand side the *control vector* or control variable.

1.5. Criterion Function

The question immediately arises about the determination of effective ways of choosing the control variable. In general, there are severe limitations on the freedom of choice that exists. These limitations, or *constraints* as they are usually called, arise both from the nature of the physical system itself and from the type of devices that are available for exerting control. Recognition of these realistic bounds on behavior usually introduces severe analytic difficulties. Sometimes these constraints present serious obstacles to a numerical solution; sometimes they greatly facilitate it. It is often a question of the method employed. For pedagogical reasons we shall defer any serious consideration of control processes of this nature until the fourth volume in this series.

Here, we will suppose that a choice of control is dictated solely by questions of *cost*. We distinguish two types of cost:

(a) the cost of exerting control,
(b) the cost of undesired behavior of the system. $\tag{1.5.1}$

These costs will presumably be determined by the behavior over time of the state and control vectors. In practice, neither type of cost is particularly easy to evaluate. We will discuss some aspects of this in Chapter 14.

To initiate the mathematical discussion, however, we will agree to measure the costs of both types by relatively simple scalar functions of x and y. Thus, for the discrete process, we take the cost of system performance to be of the form

$$\sum_{n=0}^{N-1} h_n(x_n) + k(x_N). \tag{1.5.2}$$

The function $h_n(x_n)$ can be interpreted as a cost associated with the n-th stage, and the function $k(x)$ as a cost associated with the terminal state. In particular cases, either of these costs may be taken to be zero. Similarly, we take the cost of control to have the form

$$\sum_{n=0}^{N-1} g_n(y_n). \tag{1.5.3}$$

We now make the fundamental assumption that these two kinds of costs are commensurable, i.e., measured in the same units. This assumption (frequently and frustratingly not satisfied) enables us to obtain a total cost by merely adding the two expressions in (1.5.2) and (1.5.3). The result is

$$J(\{x_n, y_n\}) = \sum_{n=0}^{N-1} [h_n(x_n) + g_n(y_n)] + k(x_N) \tag{1.5.4}$$

This function of the states and controls is called a *criterion function*. In some few cases the nature of the underlying process dictates the choice of a criterion. In general, we have a certain amount of flexibility.

Quite analogously, in an examination of continuous processes we are led to consider a *criterion functional* of the form

$$J(x, y) = \int_0^T [h(x, t) + g(y, t)] \, dt + k(x(T)). \tag{1.5.5}$$

Our excuses for treating relatively simple functions and functionals are, first, that fortuitously a number of important processes can be meaningfully treated in this fashion and, secondly, that we can often use the simple case to treat more complex cases by steadfastly using the method of successive approximations.

1.6. Control Processes

We are now prepared to precisely define a *control process*. By this term we mean the determination of the control vector which minimizes the criterion function, or functional, subject to the condition that the state and control vectors are related, as in Section 1.4.

The mathematical problems are now well-defined. As we shall see, we can employ calculus (the calculus of variations) and dynamic programming to study them. We will discuss in detail the use of the theories of the calculus of variations and dynamic programming as analytic and computational t ⊖ ls. Our aim throughout is to provide a broad basis for more intensive studies by avoiding any undue focusing on particular areas.

As might be expected, the development of the digital computer has made a number of other analytic approaches feasible, and in many cases, far more desirable. References to a number of these alternative methods will be given at appropriate points in the text. The field of optimization and control has grown so rapidly that we can do little more than provide references to some significant areas, if we do not wish the book to assume formidable proportions. Numerous important developments must necessarily be omitted from the body of the text in order to concentrate on a few techniques.

1.7. Analytic Aspects

The first volume was devoted to a detailed study of the fundamental case where the describing equations were linear and the criteria were quadratic. Thus, for example, in the discrete case, a typical problem was that of minimizing the quadratic form

$$J(\{x_n, y_n\}) = \sum_{n=0}^{N-1} [(x_n, Ax_n) + (y_n, By_n)] \tag{1.7.1}$$

over the y_n where $\{x_n\}$ and $\{y_n\}$ are connected by the relation

$$x_{n+1} = Cx_n + Dy_n, \qquad x_0 = c. \tag{1.7.2}$$

In the continuous case, a characteristic problem was that of minimizing the quadratic functional

$$J(x, y) = \int_0^T [(x, Ax) + (y, By)] \, dt, \tag{1.7.3}$$

where x and y are connected by the linear differential equation

$$x' = Cx + Dy, \qquad x(0) = c.$$

Here, we are using the inner product notation

$$(x, y) = \sum_{i=1}^{N} x_i y_i, \tag{1.7.4}$$

where now x_i and y_i are, respectively, the ith components of the M-dimensional vectors x and y.

Using either approach (the calculus of variations or dynamic programming), we were able to obtain an explicit analytic solution to these questions. These representations can be made the basis of effective computational solutions, albeit not without some thought and effort in the case where the dimension of the system, M, is large. The detailed study of these solutions was a consequence of the normal human desire to dwell upon those few tasks that can be accomplished relatively easily and elegantly, and also due to the fact that these explicit results can be used to provide a firm foundation for the effective solution of more general problems, described in Section 1.5, by means of various applications of the method of successive approximations. This will be discussed in Chapters 3 and 5.

In general, the problem of Section 1.5 cannot be resolved in explicit analytic form in terms of the familiar elementary functions. The nonlinearity of the equations effectively precludes this. Hence, the questions of obtaining both analytic and numerical estimates become matters of some considerable difficulty. One can state quite flatly that the general problem will never be resolved in terms of any single well-defined method or set of methods. What is needed in any particular case is a collection of interlocking and overlapping techniques of a flexible nature, plus a great deal of hard work and a certain amount of ingenuity.

1.8. Computational Aspects

Whether we are concerned with applications, either primarily or as a stimulating source of interesting and difficult questions, the question of effective numerical solution is of paramount importance. We are thinking, naturally, in terms of the use of digital computers, hybrid computers, and man–machine combinations when we use the term "computational."

It may turn out, as we shall illustrate, that a successful numerical approach will depend upon the use of different state variables, and even a different conceptual formulation of a control process, than that in a conventional treatment. The computer has provided an undreamt of flexibility in the domain of numerical solution of problems arising in the control area. How to use this flexibility and versatility is a constant challenge to the mathematician. There have been two computer revolutions—the first in speed of execution and the second in concept.

1.9. Event versus Time Orientation

There are at least two distinct ways in which we can consider applying control to a system:

(a) exert influences in a prescribed fashion over time; and
(b) exert influences in a prescribed fashion depending upon the events that occur. $\qquad$ (1.9.1)

Referring to the descriptive equation,

$$x_{n+1} = g(x_n, y_n), \qquad x_0 = c, \tag{1.9.2}$$

this means, first of all, that we can look for a solution in the form $y_n = h(n)$, which is the approach of calculus. In the continuous case, that of the calculus of variations, we ask for a solution which is a function of time, $y(t)$. Alternatively, we can regard a solution as a relation $y_n = k_n(x_n)$, a function of time and state, which is the approach of dynamic programming.

We have discussed some of the merits and demerits of each of these approaches in Volume I, and we will present some further discussion in the pages that follow. At the moment, we wish to recall the fact that these two formulations, so different conceptually, are analytically equivalent—as long as we restrict our attention to deterministic control processes. The equivalence is a consequence of the fundamental duality of Euclidean space. We shall exploit this duality as a device for obtaining upper and lower estimates.

Since the theories are dual, we can anticipate that an adroit combination of the two will be more powerful than either alone.

1.10. Discrete versus Continuous

Which formulation of a control process should be employed: one based upon discrete time, or one in which time is allowed to be continuous? What is a more "natural" approach to a mathematical treatment of a physical control process?

There are no quick and easy answers to general questions of the foregoing nature, nor indeed are the questions in this generality meaningful. Considerable thought should be devoted to the choice of state variables and the form of the laws that govern their behavior over time. Some of the factors entering into a choice of a mathematical model are the structure of the physical process, the type and accuracy of data that is available, the type of approximate solution desired, the analytic intricacies of the model, and, almost paramount, the type of computer that is available.

Neither the discrete nor the continuous model should be considered as an approximation to the other. Both are useful approximations, but approximations only, to reality.

BIBLIOGRAPHY AND COMMENTS

For considerable expansion of the preceding discussion and many further references, the reader may wish to refer to

R. Bellman, *Adaptive Control Processes: A Guided Tour*, Princeton Univ. Press, Princeton, New Jersey, 1961.

R. Bellman, "Control Theory," *Sci. Amer.*, Sept. 1964, pp. 186–200.

R. Bellman, *Introduction to the Mathematical Theory of Control Processes*, Vol. I: *Linear Equations and Quadratic Criteria*, Academic Press, New York, 1968.

R. E. Kalman, "Algebraic Aspects of the Theory of Dynamical Systems," in *Differential Equations and Dynamical Systems* (J. K. Hale and J. P. LaSalle, eds.) Academic Press, New York, 1967, pp. 133–146.

For discussion of the background of nonlinear problems and stability theory, see

T. J. Higgins, "A Résumé of the Development and Literature of Nonlinear Control-System Theory," *Trans. ASME*, April 1957, pp. 445–453.

J. P. LaSalle, "Stability and Control," *J. SIAM Control*, 1, 1962, pp. 3–15.

T. Von Karman, "The Engineer Grapples with Nonlinear Problems," *Bull. Amer. Math. Soc.*, **46**, 1940, pp. 615–683.

P. Whittle, A View of Stochastic Control Theory, *J. Royal Stat. Soc.*, **132**, 1969, pp. 320–335.

D. M. G. Wishart, "A Survey on Control Theory," *J. Royal Stat. Soc.*, **132**, 1969, 293.

For some discussion of the many questions associated with formulation, see

R. Bellman and P. Brock, "On the Concepts of a Problem and Problem-Solving," *Amer. Math. Monthly*, **67**, 1960, pp. 119–134.

K. Borch, "Economic Objectives and Decision Problems," *IEEE Trans. Sys. Sci. and Cybernet.*, Sept. 1968, pp. 266–270.

M. W. Shelley, II and G. L. Bryan (eds.), *Human Judgments and Optimality*, Wiley, New York, 1964.

L. Tisza, "The Conceptual Structure of Physics," *Rev. Mod. Phys.*, **35**, 1963, pp. 151–185.

L. Zadeh, "What Is Optimal?" *IRE Trans. Inform. Theory*, **IT-4**, 1958, p. 3.

2

DISCRETE CONTROL PROCESSES

AND DYNAMIC PROGRAMMING

2.1.   Introduction

Let us now turn our attention to discrete control processes. We will focus initially upon the general problem of minimizing a scalar criterion function of the form

$$J(\{x_n, y_n\}) = \sum_{n=0}^{N} h_n(x_n, y_n) + \phi(x_N) \tag{2.1.1}$$

with respect to the x_n and y_n, where these vector variables are related by means of the equations

$$x_{n+1} = g(x_n, y_n), \qquad x_0 = c, \tag{2.1.2}$$

$n = 0, 1, 2, \ldots, N$.

We are primarily concerned with illustrating the use of the theory of dynamic programming to treat questions of this nature and with illustrating how this new formulation may be utilized to obtain various types of information concerning the structure of the solution.

One of the considerable advantages of beginning with the discrete case is that questions of existence of a solution are readily resolved. Furthermore, simple modifications enable us to treat the case where constraints are present and to indicate the applicability of these ideas to important classes of discrete control processes of stochastic type. We are thus in a position to tackle a number of basic problems without a great deal of preliminary effort. Consequently, from the pedagogical point of view there is considerable merit to considering discrete processes first.

In Chapter 3 we turn to a discussion of some computational aspects of the dynamic programming algorithms.

2.2. Existence of a Minimum

In the study of any variational problem the first important question is determining whether the problem is well posed, which is to say, can we guarantee that a minimum or maximum value is actually attained? Once this has been established, one can continue to the more interesting study of the qualitative properties of the minimizing or maximizing function, culminating in the most difficult task of all, the discovery of feasible means of obtaining numerical values, quantitative properties. In general, in control processes of significance, questions of existence are not as important or as difficult as questions of determination of feasible computational techniques. Consequently, in a first discussion of the type we present here, it is best to consider control processes where the existence of a solution is easy to demonstrate so that we can concentrate more fully on the analytic and computational aspects. We shall hew to this line throughout.

There are two cases of particular importance. The first is where both the state and the control spaces are discrete and finite and where we suppose that the criterion function is defined for all possible sets of state and control vectors. It follows that the minimization problem involves the search for the smallest of a finite set of values, and thus necessarily possesses a solution.

The second case of importance is that where $g(x, y)$, and more generally $J(\{x_n, y_n\})$, is continuous in its arguments over the entire x and y space. If we can establish the fact that the domains over which the state and control vectors range are closed and bounded, then it will again follow that the minimum value is attained. Sometimes, we constrain these domains from the beginning; sometimes, we can establish a priori bounds which accomplish the same purpose.

Let us give an example of this. Suppose that

$$J(x, y) = \sum_{n=0}^{N} [(x_n, x_n) + (y_n, y_n)], \qquad (2.2.1)$$

where, as before,

$$x_{n+1} = g(x_n, y_n), \qquad x_0 = c, \qquad (2.2.2)$$

and we wish to determine the minimum of J over all $\{y_n\}$.

Suppose that we consider the possible control values

$$y_0^{(0)} = y_1^{(0)} = \cdots = y_{N-1}^{(0)} = 0 \qquad (2.2.3)$$

with the corresponding values of the x_n determined by the relation

$$x_{n+1}^{(0)} = g(x_n^{(0)}, 0), \qquad x_0^{(0)} = c, \qquad n = 0, 1, 2, \ldots, N-1. \qquad (2.2.4)$$

Since we are interested in the absolute minimum of J, it follows that we need consider only state and control vectors satisfying the constraint

$$\sum_{n=0}^{N} [(x_n, x_n) + (y_n, y_n)] \le \sum_{n=0}^{N} (x_n^{(0)}, x_n^{(0)}). \qquad (2.2.5)$$

This determines a bounded, closed region in x_n and y_n space. The assumption that g is continuous is then enough to ensure the existence of a set of $\{x_n\}$ and $\{y_n\}$ which minimize.

This is an important idea in the theory of control processes. Any admissible choice of control variables automatically yields an *upper* bound for J. If we could as easily determine lower bounds, we might by judicious choices obtain quite accurate estimates for the minimum value of J. This will be the theme of Chapter 9.

2.3.   Uniqueness

The problem of uniqueness of the minimizing control vectors is much more difficult to handle. Generally, we must impose some convexity conditions to ensure this property, which plays such a major role in determining the structure of the solution.

Fortunately, as we shall point out, the algorithms of dynamic programming automatically yield the absolute minimum even when other relative minima exist. Hence, as far as a computational solution is concerned, convexity is not as vital in the theory of dynamic programming as it is in other theories of optimization. Nonetheless, whenever this property is present, we can always use it to good avail.

2.4.   Dynamic Programming Approach

Returning to the minimization problem posed in Section 2.1, let us agree to regard it as a multistage decision process in which at each stage the control variable y_n is to be chosen in terms of the current state, x_n, and the number of stages remaining in the process. We are abstracting the mathematical content of the fundamental engineering principle of *feedback control*. The influence exerted upon the system is dependent upon the current state of the system. This is an example of event orientation.

Suppose that we start in state $x_0 = c$ and choose the control vector y_0, as in Figure 2.1. Referring to (2.1.1), we see that the contribution to the criterion function J, as a result of this decision y_0, is $h(x_0, y_0)$. We shall begin with the case where the return per stage is independent of time. Thus, $h_0, h_1, \ldots h_{N-1}$ are all equal to h.

Figure 2.1

Starting in the new state x_1, we choose y_1; see Figure 2.2. The new contribution to J, is $h(x_1, y_1)$.

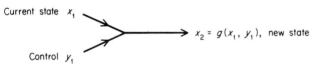

Figure 2.2

We observe then that this may be considered to be a multistage decision process with an important invariant structure. Due to the additivity of the criterion function in (2.1.1) and the nature of the recurrence relation in (2.1.2), at each stage we are in a position where we can ignore the past (both the past history of the process and the past costs) when it comes to making present and future decisions. We are free to concentrate on choosing the remaining control variables, so as to minimize the remaining part of the criterion function, by starting with a knowledge of the current state.

It is this sharp decomposition into a past, present and future which permits us to obtain an analytic hold on the optimization process Time is playing its customary fundamental role of stratification.

2.5. Recurrence Relations

The foregoing observations can readily be translated into an analytic treatment of the minimization problem which is quite distinct from the usual approach of calculus It permits us to replace a multidimensional variational problem by a sequence of lower-dimensional problems. We begin with the obvious comment that the minimum value of J depends upon the initial value c and the number of stages N. Hence, we can write

$$\min_{\{x_n, y_n\}} J(\{x_n, y_n\}) = f_N(c). \tag{2.5.1}$$

Let us emphasize that this is an *absolute minimum*, not merely a stationary value. This function $f_N(c)$ is defined for $N = 0, 1, 2, \ldots,$ and for c in the domain of the state variable. At the moment we allow all values of c, and likewise a free variation for y.

We have

$$f_0(c) = \min_{y_0} [h(c, y_0) + \phi(c)] \tag{2.5.2}$$

To obtain a relation between f_N and f_{N-1}, we use a simple basic functional property of the minimum operation, namely that

$$\min_{[y_0, y_1, \ldots, y_N]} = \min_{y_0} \min_{[y_1, y_2, \ldots, y_N]} \tag{2.5.3}$$

Hence,

$$f_N(c) = \min_{[y_0, y_1, \ldots, y_N]} [\cdots] = \min_{y_0} \min_{[y_1, y_2, \ldots, y_N]} [\cdots]$$

$$= \min_{y_0} \left[h(c, y_0) + \min_{[y_1, y_2, \ldots, y_N]} [h(x_1, y_1) + \cdots + \phi(x_N)] \right]. \tag{2.5.4}$$

To simplify the second term in the bracket on the right, we observe that it corresponds to the value of the criterion function for an $(N - 1)$-stage process starting in the state $x_1 = g(x_0, y_0) = g(c, y_0)$. Hence, using the function defined in (2.5.1) for $N = 1, 2, \ldots,$ and all c, we can write

$$f_N(c) = \min_{y_0} [h(c, y_0) + f_{N-1}(g(c, y_0))], \tag{2.5.5}$$

the desired recurrence relation. Observe that y_0 is now a dummy variable and we can replace it by y to simplify the typography.

2.6. Imbedding

The classical approach to optimization processes of the foregoing type is to consider c and N as fixed and to determine the set of variables $[y_0, y_1, \ldots, y_N]$, which provide the minimum of $J(\{x_n, y_n\})$; see Figure 2.3. This procedure traces out a single trajectory in phase space, the points $[x_0, x_1, \ldots, x_N]$, the optimal trajectory. There are many natural advantages to this point of view, as we shall see in Chapter 4 when we consider continuous control processes.

The approach of dynamic programming pursued above regards c and N as *state variables*. In place of attempting to solve a particular problem where c and N are fixed, we are attacking a family of problems obtained by allowing c and N to assume all admissible values. This is an *imbedding* procedure.

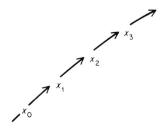

Figure   2.3

In place of a single trajectory from a fixed point, we determine the optimal trajectory from any point in the state variable domain. In many situations this family of solutions is what is desired from the very beginning. In other situations, we accept this plethora of information as the price of obtaining the particular solution for which we are looking.

2.7.   Policies

The recurrence relation of (2.5.5) determines two functions. First of all, it determines the sequence of absolute minima, $\{f_N(c)\}$. Secondly, it determines the function $y_0(c)$ which furnishes the absolute minimum in (2.5.5).

Observe that there are some difficulties of notation here. The function $y_0(c)$, the function yielding the minimum in (2.5.5), should perhaps be written in the form $y_{0N}(c)$ to indicate the fact that it is the decision made in state c when there are N stages remaining. This would be consistent with the notation $f_N(c)$. On the other hand, multiple subscripts are a nuisance. Let us compromise and write $y(c, N)$ to avoid confusion with the control variable $y_N(c)$.

In general, we call a function $y(c, N)$ which yields an admissible choice of y_0 a *policy*. A policy which yields the absolute minimum of the criterion function is called the *optimal policy*. Observe that we can consider policies even when there is no criterion function.

We see that the function $f_N(c)$ is uniquely determined by (2.5.5). However, the optimal policy, $y(c, N)$, need not be a single-valued function since the minimum in (2.5.5) can be attained by several different values of y_0. A knowledge of $f_N(c)$ allows us to find all values of $y(c, N)$. Conversely, any optimal policy determines $f_N(c)$.

Policy is a basic concept of dynamic programming. A systematic investigation of the ramifications of this concept enables us both to extend the scope of classical analysis and to formulate many novel types of problems and problem areas.

2.8.   Principle of Optimality

In Section 2.5 we derived the basic recurrence relation (2.5.5), directly from the functional properties of the minimum operation. For future purposes let us observe that we were applying a fundamental result concerning the structure of optimal policies:

PRINCIPLE OF OPTIMALITY. *An optimal policy has the property that whatever the initial state and the initial decision are, the remaining decisions must constitute an optimal policy with regard to the state resulting from the first decision.*

In most cases, the application of this principle is routine and the validity immediate. In some cases, however, it is necessary to choose the state variables carefully and to investigate carefully the requisite independence of the future from the past.

2.9.   Discussion

Intuitively, we can think of the determination of optimal control as a problem of tracing out a geodesic in some appropriate phase space. Let P be the initial point and Q the terminal point of the path, and suppose initially that we want a path of least time from P to Q; see Figure 2.4. It is clear then that

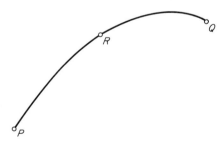

Figure   2.4

any particular tail of this curve, the part from any point R on the path to Q, must also be a path of least time. This is an illustration of the principle of optimality.

But what about the part of the path from P to R? Is that not also a path of minimal time? Can we not extend the principle enunciated in Section 2.8 to include any intermediate set of decisions, or conversely, can we conclude that the principle of optimality is merely a consequence of the foregoing simple observation? The answer is "No!"

The basic point is that the simple optimization process formulated above possesses special homogeneity features which are not at all characteristic of the general situation. Suppose that we slightly modify the foregoing problem to read: Determine a path from P to Q which arrives at Q from a specified direction in minimum time. Then it is clear as before that RQ is a path of the same type starting from R. But PR is not necessarily a path of the same type, and, in general, will not be.

Once R has been specified, PR certainly possesses an optimal property, but it is not clear, a priori, how to specify this optimality. On the other hand, RQ is an optimal path with respect to the original criterion. It is this invariance property that is fundamental, and as we see from (2.5.5), a determination of PR ultimately hinges upon this fact.

2.10.   Time Dependence

So far we have used the homogeneity in time implicit in the relation

$$x_{n+1} = g(x_n, y_n), \qquad x_0 = c, \tag{2.10.1}$$

and in our assumption that the single-stage return functions were independent of time. Suppose that the transformation of states is time-dependent, which is to say, suppose that it has the form

$$x_{n+1} = g_n(x_n, y_n), \qquad x_0 = c. \tag{2.10.2}$$

In that case, we reverse the time-axis to obtain the desired invariance Introduce the function

$$\phi_k(c) = \min_{\{y_n\}} \left[\sum_{n=k}^{N} h_n(x_n, y_n) + \phi(x_N) \right]. \tag{2.10.3}$$

We can proceed as before to obtain a recurrence relation.

2.11.   Constraints

In many control processes of interest, an unrestricted choice of a control action is not possible. The constraints on the control variable are either of economic, engineering, or physical nature and take a variety of forms. To

make a restriction of this type mathematically precise, we shall suppose here
that y, the control variable, is allowed only to range over a portion of the
entire decision space D and that this portion is determined by the current
state and the stage. We write

$$y \in D_N(c) \tag{2.11.1}$$

to indicate this. In place of (2.5.5), we obtain the functional equation

$$f_N(c) = \min_{y \in D_N(c)} [h(c, y) + f_{N-1}(g(c, y))]. \tag{2.11.2}$$

The presence of the constraint introduces a number of analytic intricacies,
as is to be expected. Nonetheless, as we shall show in Section 2.16 et seq., the
functional equation can be used to determine the structure of the optimal
policy in a number of cases. Further examples will be found in the exercises.
What is surprising at first glance, however, is the fact that the presence of
constraints considerably simplifies the computational aspects, which we shall
discuss in Chapter 3.

There are two types of constraints of particular importance, *local* and
global. Local constraints are restrictions either directly in the choice of control
variable or indirectly by allowing only a specified set of states, $S_k(c)$, to which
one can go from a particular state c.

Global constraints are restrictions on the entire set of control and state
variables, essentially on the history of the process. Thus a global constraint
might have the form

$$\sum_{n=0}^{N} k_n(x_n, y_n) \le a_1. \tag{2.11.3}$$

In many important cases they occur in this separable form.

As we shall see a few pages further on, global constraints can often be
accounted for in a simple and meaningful fashion.

2.12.   Analytic Aspects

The question arises as to what can actually be done with functional equations of the type appearing in (2.5.5) and (2.11.2). There are a number of
possible objectives:

(a)   Determine the analytic form of $f_N(c)$ in terms of the analytic nature of
   g, h, and $D_N(c)$.
(b)   Determine the structure of the optimal policy.
(c)   Determine the asymptotic behavior of $f_N(c)$ and $y(c, N)$ as $N \to \infty$.
(d)   Determine methods for deriving convenient analytic approximations
   for $f_N(c)$ and $y(c, N)$. In particular, obtain upper and lower bounds.

(e) Introduce functional transformations which determine $f_N(c)$ simply and explicitly in carefully chosen cases.

(f) Consider inverse problems, i.e., given $y(c, N)$, what classes of functions g and h possess optimal policies of this nature?

(g) Consider the use of (2.5.5) and (2.11.2) to determine numerical values of $f_N(c)$ and $y(c, N)$.

This is a broad problem and we cannot possibly hope to explore all aspects in this volume. In Chapter 3, we discuss some of the questions involved in the use of (2.5.5) and (2.11.2) for numerical purposes. In the following sections we consider examples of the types of questions proposed above, and in the exercises we provide some further illustrations of directions of research suggested by dynamic programming. Extensive additional references will be found at the end of this chapter.

2.13. Marginal Returns

Let us begin with an important general result which is useful in many cases. We consider the scalar version of (2.5.5)

$$f_N(c) = \min_v \; [h(c, v) + f_{N-1}(g(c,v))]. \tag{2.13.1}$$

The minimizing v may be determined as the solution of

$$0 = h_v + g_v f'_{N-1}(g(c, v)), \tag{2.13.2}$$

if we suppose that it is permissible to use calculus, as we do. Then (2.13.1) yields the relation

$$f_N(c) = h(c, v) + f_{N-1}(g(c, v)) \tag{2.13.3}$$

with v determined by (2.13.2). Differentiating both sides of (2.13.3) with respect to c, we have

$$\begin{aligned} f'_N(c) &= h_c + g_c f'_{N-1}(g(c, v)) + [h_v + g_v f'_{N-1}(g(c, v))]v_c \\ &= h_c + g_c f'_{N-1}(g(c, v)). \end{aligned} \tag{2.13.4}$$

If we call $f_N(c)$ the *marginal return* function, we can determine the optimal policy in terms of the marginal return function, and the marginal function itself recurrently, using (2.13.2) and (2.13.4) without needing the return function.

EXERCISE

1. Obtain the multidimensional version of the foregoing results.

2.14. Linear Equations and Quadratic Criteria

The most important case in which analytic solutions can be readily obtained is that where the descriptive equation is linear,

$$x_{n+1} = Ax_n + y_n, \qquad x_0 = c, \tag{2.14.1}$$

and the criterion function is quadratic, say

$$J[\{x_n, y_n\}] = \sum_{n=0}^{N} [(x_n, x_n) + (y_n, y_n)]. \tag{2.14.2}$$

We devoted Volume I to a detailed examination of processes of this structural form, considering both the discrete and continuous versions. Let us here quickly review the basic results for the discrete process defined by (2.14.1) and (2.14.2). This review is useful, as well as expository, since we wish to use results of this nature as the basis of various approximation techniques in this and later chapters. Furthermore, we may consider the results to be immediate applications of the results of Section 2.13.

We will consider only the specific problems above, reserving some more complex versions for the exercises. Write, as usual,

$$f_N(c) = \min_{\{y\}} J(\{x_n, y_n\}), \tag{2.14.3}$$

defined for $N = 0, 1, 2, \ldots$, and all c. The fundamental recurrence relation of (2.5.5) yields

$$f_N(c) = \min_{y} [(c, c) + (y, y) + f_{N-1}(Ac + y)], \tag{2.14.4}$$

with $f_0(c) = (c, c)$.

It is inductively clear from this relation, or by using the linear variational equation associated with the variational problem posed by (2.14.1) and (2.14.2) that $f_N(c)$ is a quadratic form in c,

$$f_N(c) = (c, R_N c), \tag{2.14.5}$$

where R_N is a positive definite symmetric matrix dependent only on N. To obtain a recurrence relation for R_N in terms of R_{N-1}, we use (2.14.4) obtaining

$$(c, R_N c) = \min_{y} [(c, c) + (y, y) + (Ac + y, R_{N-1}(Ac + y)). \tag{2.14.6}$$

The variational equation for y is

$$y + R_{N-1}Ac + R_{N-1}y = 0, \tag{2.14.7}$$

$$y = -(I + R_{N-1})^{-1} R_{N-1}Ac.$$

Rather than substitute this directly in (2.14.6), we observe that (2.14.7) yields

$$(y, y) + (y, R_{N-1}(Ac + y)) = 0. \tag{2.14.8}$$

Using this relation in (2.14.6), we have

$$(c, R_N c) = (c, c) + (Ac, R_{N-1}(Ac + y)), \tag{2.14.9}$$

and thus

$$(c, R_N c) = (c, c) - (Ac, y), \tag{2.14.10}$$

using (2.14.7) again. Hence,

$$(c, R_N c) = (c, c) + (Ac, (I + R_{N-1})^{-1} R_{N-1} Ac),$$

yielding for $N \geq 1$ the recurrence relation

$$R_N = I + A^T (I + R_{N-1})^{-1} R_{N-1} A$$

with $R_0 = I$. Here A^T denotes the transpose of A. We deviate from the preferable prime notation, A', to avoid possible confusion with the derivative.

Preliminary simplifications of the foregoing type, taking advantage of the variational equation, will often obviate some messy matrix manipulation.

EXERCISES

1. Consider the asymptotic behavior of the return and control functions as $N \to \infty$, first in the scalar case and then in the multidimensional case.
2. Consider the case where the criterion function has the form $\sum_{n=0}^{N} [(x_n, Cx_n) + (x_n, Dy_n) + (y_n, Ey_n)]$ and the describing equation is $x_{n+1} = Ax_n + By_n$, $x_0 = c$. Discuss separately the cases where B is singular and nonsingular.
3. Obtain results corresponding to the foregoing when A, B, C, D, E are time-dependent.
4. Consider the case where the criterion function has the form $\sum_{n=0}^{N} [(x_n, x_n) + (y_n, y_n) + \lambda(x_N - a, x_N - a)$ with $\lambda \geq 0$. What are the limiting forms of the control vector and the return function as $\lambda \to \infty$?
5. Consider the problem of minimizing $J(\{u_n, v_n\}) = \sum_{n=0}^{N} [u_n^2 + u_n^4 + v_n^2]$ where $u_{n+1} = au_n + v_n$, $u_0 = b$. Write $f_N(b) = r_N b^2 + r_{1N} b^4 + \cdots$. From the functional equation $f_N(b) = \min_v [b^2 + b^4 + v^2 + f_{N-1}(ab + v)]$, obtain recurrence relations for r_N, r_{1N}, ... Does the series for $f_N(b)$ above possess a nonzero radius of convergence?

6. Extend to the multidimensional case.

7. Consider the problem of minimizing $J(\{u_n, v_n\}) = \sum_{n=0}^{N} [u_n{}^2 + \varepsilon u_n{}^4 + v_n{}^2]$ where $u_{n+1} = au_n + v_n$, $u_0 = b$, and $\varepsilon \geq 0$. Write $f_N(b) = \min J = f_{0N}(b) + \varepsilon f_{1N}(b) + \cdots$, and obtain recurrence relations for $f_{0N}, f_{1N}, \dots$ Does the power series for $f_N(b)$ possess a nonzero radius of convergence

8. despite the fact that J does not possess an absolute minimum for $\varepsilon < 0$?

2.15. Discussion

Returning to (2.14.7) we see that the optimal policy, $y(c, N)$, has a particularly simple form, namely

$$y(c, N) = -(I + R_{N-1})^{-1} R_{N-1} Ac = S_N c, \qquad (2.15.1)$$

where S_N is independent of c. The feedback control is linear, which is to say, the control variable is a linear transformation of the state variable.

This leads to an interesting observation. Suppose that we observe linear feedback control in a physical process. We can then conclude that the control process may be regarded as a multistage decision process in which the decisions are being made *as if* there existed a quadratic criterion function.

This in turn leads to an interesting class of problems which are usually called "inverse problems." Given an observed behavior,

$$y = \phi(x), \qquad (2.15.2)$$

which is to say, a situation in which the response is determined by the stimulus, can we construct a control process in which this represents an optimal policy? We shall have more to say about this in Chapter 6 where we discuss interconnections between dynamic programming and the calculus of variations.

In general, an explicit analytic solution to a control process cannot be expected, and one has to be satisfied with a description of the structures of the optimal policy and the return function. In many cases, fortunately, this is what is really desired. In the following sections we will give some examples of analytic results that can be obtained from a knowledge of simple structural properties of the functions appearing in the describing equations and in the criterion function.

EXERCISE

1. Given an observed control law $y = Bc$, what quadratic criteria and describing functions can we associate with this feedback process?

2.16. An Example of Constraints

To illustrate the kinds of results that can be obtained where constraints are present and the kinds of arguments that can be used, consider the following scalar control process. Let

(a) $u_{k+1} = u_k + b(v_k), \quad u_0 = c,$

(b) $0 \le v_k \le u_k, \quad k = 0, 1, 2, \ldots, N,$ (2.16.1)

(c) $J(\{u_k, v_k\}) = \sum_{k=0}^{N} (u_k - v_k).$

Setting

$$f_N(c) = \max_{\{v_k\}} J(\{u_k, v_k\}),$$ (2.16.2)

we readily obtain the recurrence relations

$$f_{N+1}(c) = \max_{0 \le v \le c} [c - v + f_N(c + b(v))], \quad N = 0, 1, 2, \ldots,$$ (2.16.3)

setting $f_0(c) = c$.

Our aim is to determine the structure of the optimal policy, $v_N(c)$, under some simple assumptions concerning $b(v)$. Let us assume

(a) $b(0) = 0, \quad b'(0) = \infty,$

(b) $b'(v) > 0, \quad b'(v) \to 0 \quad \text{as} \quad v \to \infty,$ (2.16.4)

(c) $b''(v) < 0.$

A simple function satisfying all of these requirements is $b(v) = v^{1/2}$. Geometrically, we are thinking of a concave function of the type shown in Figure 2.5.

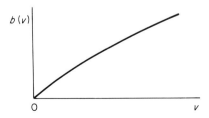

Figure 2.5

We wish to demonstrate that under the foregoing assumptions the optimal policy has the following structure:

(a) $v_N(c) = c,$ $0 \le c \le c_N,$ (2.16.5)

(b) $0 < v_N(c) < c,$ $c_N < c,$

where the c_i are a sequence of numbers which are monotone increasing, $c_1 < c_2 < \cdots$. We will indicate below how these numbers are determined.

The derivation of results will be inductive. Let us begin with the case $N = 1$. We have

$$f_1(c) = \max_{0 \le v \le c} [c - v + f_0(c + b(v))].$$ (2.16.6)

If an internal maximum occurs, it occurs at a point where

$$1 = b'(v)f_0'(c + b(v)),$$ (2.16.7)

which is to say $1 = b'(v)$, since $f_0(c) = c$.

By virtue of the assumptions made in (2.16.4), we can assert that there is precisely one solution to this equation; call it c_1.

If $c \le c_1$, the maximum in (2.16.6) will occur at $v = c$. If $c > c_1$, there will be an internal maximum. Hence,

$$f_1(c) = f_0(c + b(c)),   c \le c_1,   v = c,$$

$$= c - v_1 + f_0(c + b(v_1)),   c > c_1,   v = v_1 < c.   (2.16.8)$$

The function $f_1(c)$ is clearly differentiable for $c \le c_1$ and for $c > c_1$. We have

$$f_1'(c) = [1 + b'(c)]f_0'(c + b(c)) = 1 + b'(c),   c \le c_1,$$

$$= 1 + f_0'(c + b(v_1)) = 2,   c > c_1.   (2.16.9)$$

Since $b'(c_1) = 1$ at $c = c_1$, we see that $f_1(c)$ has a continuous derivative for $c \ge 0$. Furthermore, $f_1'(c) > f_0'(c)$ for $c > 0$.

Let us now investigate concavity. We have

$$f_1''(c) = b''(c),   c \le c_1,$$

$$= 0,   c > c_1.   (2.16.10)$$

Hence $f_1''(c) \le 0$ for all $c \ge 0$.

Consider next the case $N = 2$. We have

$$f_2(c) = \max_{0 \le v \le c} [c - v + f_1(c + b(v))].$$ (2.16.11)

If an internal maximum occurs, it occurs at a point where

$$\frac{1}{b'(v)} = f_1'(c + b(v)).$$ (2.16.12)

Since $b'(v) > 0$, $f''_1(c + b(v)) \le 0$, $(1/b'(v))' > 0$, with $1/b'(0) = 0$, $1/b'(\infty) = \infty$, there is exactly one root, which we call $v_2 = v_2(c)$. Since $f'_1 > f'_0$, it follows that $v_2(c) > v_1(c)$.

The critical value of c is the root of

$$\frac{1}{b'(c)} = f'_1(c + b(c)). \tag{2.16.13}$$

Call this root c_2. Since $f'_1 > f'_0$ and $(1/b')' > 0$, we have $c_2 > c_1$.

Thus,

$$f_2(c) = f_1(c + b(c)), \qquad\qquad 0 \le c \le c_2,$$
$$= c - v_2 + f_1(c + b(v_2)), c \ge c_2. \tag{2.16.14}$$

A direct calculation of $f'_2(c)$ inside the two intervals of interest shows that $f'_2(c)$ is continuous at c_2.

Let us now turn to the concavity of $f_2(c)$. We have

$$f''_2(c) = b''(c)f'_1(c + b(c)) + [1 + b'(c)]f''_1(c + b(c)), \qquad c \le c_2,$$
$$= f''_1(c + b(v_2))[1 + b'(v_2)v'_2(c)], \qquad c > c_2. \tag{2.16.15}$$

To obtain an expression for $v'_2(c)$, we use (2.16.13) and differentiate,

$$\frac{-b''(v_2)}{b'(v_2)^2} v'_2(c) = f''_1(c + b(v_2))[1 + b'(v_2)v'_2(c)]. \tag{2.16.16}$$

Hence,

$$v'_2(c)\left[b'(v_2)f''_1(c + b(v_2)) + \frac{b''(v_2)}{b'(v_2)^2}\right] = -f''_1(c + b(v_2)), \quad (2.16.17)$$

and thus

$$v'_2(c) < 0. \tag{2.16.18}$$

Returning to (2.16.16), we see that this last result, plus $f''_1 \le 0$, implies $1 + b'(v_2)v'_2(c) > 0$. Since $f''_1 \le 0$, this yields from (2.16.15), the result $f''_2(c) \le 0$ for $c > c_2$. The result for $c < c_2$ follows from the first relation in (2.16.15). The function $f''_2(c)$ is not continuous at $c = c_2$.

The final step is to show that $f'_2(c) > f'_1(c)$. With this established, we have all the ingredients of an indirect proof. The proof depends upon a consideration of three distinct intervals $[0, c_1]$, $[c_1, c_2]$, and $[c_2, \infty]$. In $[0, c_1]$, we have

$$f'_2(c) = [1 + b'(c)]f'_1(c + b(c)) > [1 + b'(c)]f'_0(c + b(c)) = f'_1(c). \quad (2.16.19)$$

In $[c_2, \infty]$, we have

$$f'_2(c) = 1 + \frac{1}{b'(v_2)} > 1 + \frac{1}{b'(v_1)} = f'_1(c), \tag{2.16.20}$$

since $v_2 > v_1$. The remaining interval is $[c_1, c_2]$ in which we have

$$f_1'(c) = 1 + \frac{1}{b'(v_1)}, \qquad f_2'(c) = [1 + b'(c)]f_1'(c + b(c)). \quad (2.16.21)$$

Since $v_2 = c$ in $[c_1, c_2]$, we have

$$-1 + b'(v)f_1'(c + b(v)) \geq 0 \quad \text{for } 0 \leq v \leq c_2. \quad (2.16.22)$$

Hence,

$$b'(c)f_1'(c + b(c)) \geq 1. \quad (2.16.23)$$

It remains to show that

$$f_1'(c + b(c)) \geq \frac{1}{b'(c)}, \qquad c_1 \leq c \leq c_2. \quad (2.16.24)$$

Since $1/b'(v)$ is increasing and $v_1(c) \leq c$, we have $1/b'(v_1) \leq 1/b'(c)$. Using (2.16.23) we see that we have

$$f_1'(c + b(c)) \geq \frac{1}{b'(c)} \geq \frac{1}{b'(v_1)}, \qquad c \leq c_2. \quad (2.16.25)$$

2.17.   Summary of Results

The foregoing analysis provides the basis for an inductive proof of the following results: For each N, there exists a function $v_N(c)$ defined for $c \geq 0$ with the following properties:

(a) $v_N(c)$ is monotone decreasing as c increases.
(b) $v_{N+1}(c) > v_N(c)$, $N = 1, 2$.
(c)   There is a unique solution of $v_N(c) = c$, called c_N; $c_{N+1} > c_N$.
(d)   For $0 \leq c \leq c_N$, we have $f_N(c) = f_{N-1}(c + b(c))$.
(e)   For $c_N \leq c$, we have $f_N(c) = c - v_N(c) + f_{N-1}(c + b(v_N(c)))$.
(f) $f_N'(c) \geq f_{N-1}'(c)$, $N = 1, 2, \ldots, c \geq 0$.
(g) $f_N''(c) \leq 0$, $N = 1, 2, \ldots, c \geq 0$.

EXERCISES

1.   Consider the problem of minimizing the expression

$$J(\{u_k, v_k\}) = \sum_{k=0}^{N} (u_k^2 + v_k^2),$$

where $u_{k+1} = au_k + v_k$, $u_0 = c$, and we impose constraints of one of the following types:

(a) $|v_k| \leq M$,

(b) $|v_k| \leq M|u_k|$,

(c) $|v_k| \leq m_k|u_k|$, $k = 0, 1, \ldots, N$.

2.   Consider the problem of minimizing $J(\{u_k, v_k\}) = \sum_{k=1}^{N}(|u_k| + |v_k|)$ under the same types of conditions.

2.18.   Lagrange Parameter

Let us consider the scalar control process defined by

(a) $u_{k+1} = g(u_k, v_k)$, $u_0 = c$,

$$(b)   J(\{u_k, v_k\}) = \sum_{k=0}^{N} h(u_k, v_k),$$ $\quad(2.18.1)$

and let us suppose that there is a "resource constraint" of the form

$$\sum_{k=0}^{N} v_k = b,$$ $\quad(2.18.2)$

with a nonnegativity condition

$$v_k \geq 0,   \quad k = 0, 1, \ldots, N - 1.$$ $\quad(2.18.3)$

One way to treat this problem is to regard b as an additional state variable. Write

$$f_N(c, b) = \min_{\{u_k, v_k\}} J,$$ $\quad(2.18.4)$

defined for $b \geq 0$, $-\infty < c < \infty$, $N = 0, 1, 2, \ldots$, Then

$$f_0(c, b) = h(c, b),$$ $\quad(2.18.5)$

and

$$f_N(c, b) = \min_{0 \leq v \leq b} [h(c, v) + f_{N-1}(g(c, v), b - v)],   \quad N \geq 1.$$ $\quad(2.18.6)$

Since there are analytic and computational objections to functions of two variables as opposed to functions of one variable, let us see if there is a way of reducing the dimension of state space. A standard way of doing this is to introduce a Lagrange multiplier. As we shall see, this substitutes a "price" for the use of a resource as opposed to a constraint on the total quantity available.

Consider the new problem of minimizing

$$J(\{u_k, v_k\}, \lambda) = \sum_{k=0}^{N} h(u_k, v_k) + \lambda \sum_{k=0}^{N} v_k, \qquad (2.18.7)$$

subject now only to the constraint $v_k \geq 0$. Here λ is the *Lagrange multiplier*. Then, setting

$$g_N(c, \lambda) = \min_{\{u_k, v_k\}} J(\{u_k, v_k\}, \lambda), \qquad (2.18.8)$$

we have

$$g_N(c, \lambda) = \min_{v \geq 0} [h(c, v) + \lambda v + g_{N-1}(g(c, v), \lambda)]. \qquad (2.18.9)$$

We solve this problem for a range of values of λ and then attempt to adjust the value of λ until the constraint of (2.18.2) is satisfied. Presumably, this solves the original problem.

2.19.   Discussion

A rigorous treatment of the foregoing method is not easy. Some important results can, however, be quite readily obtained. The first is as follows:

(a)   If for a particular value of λ, say $\bar{\lambda}$, we find values of the v_k, say $v_k(\bar{\lambda})$, which minimize $J(\{u_k, v_k\}, \bar{\lambda})$, such that $\sum_{k=0}^{N} v_k(\bar{\lambda}) = b$, then these values minimize $J(\{u_k, v_k\})$ in (2.18.16) subject to (2.18.2).

The second is:

(b)   As λ increases, the total quantity of resources used, $\sum_{k=0}^{N} v_k(\lambda)$ decreases.

What we would like to assert is that the function

$$b(\lambda) = \sum_{k=0}^{N} v_k(\lambda) \qquad (2.19.1)$$

is continuous as λ varies from 0 to ∞. This would guarantee the existence of a solution for b in a specified range, $0 \leq b \leq b_{\max}$, where $b_{\max}$ could of course be infinite. Unfortunately, this is not always the case.

The results (a) and (b) were demonstrated in Volume I. Their proof is not difficult, and we suggest that the reader attempt them as exercises before referring to the proofs in Volume I.

EXERCISES

1. Examine the validity of minimizing $\sum_{i=1}^{N} u_i^2$ subject to the constraint $\sum_{i=1}^{N} u_i = a$, $u_i \geq 0$.
2. What about the corresponding maximization problem?
3. Examine the validity of minimizing $\sum_{i=1}^{N} u_i^2$ subject to the constraint $\sum_{i=1}^{N} u_i = a_1$, $\sum_{i=1}^{N} b_i u_i = a_2$.

2.20. Courant Parameter

Consider the problem posed in (2.18.1) with an end-point condition

$$u_N = a. \tag{2.20.1}$$

In many cases, as we shall particularly see in the continuous case in Chapter 5, a terminal constraint of this type creates some difficulties. Let us then replace this control process by the problem of minimizing

$$J(\{u_k, v_k\}, \lambda) = \sum_{k=0}^{N} h(u_k, v_k) + \lambda(u_N - a)^2, \tag{2.20.2}$$

subject to (2.18.1a). The quantity λ appearing above is called a *Courant parameter*.

We expect that the limit of the solution of this problem as $\lambda \to \infty$ is the solution to the original problem, provided that there exists a set of controls, $\{v_k\}$, for which (2.20.1) is satisfied. References to a detailed treatment of these matters will be found at the end of the chapter.

EXERCISES

1. As λ increases, does $(u_N(\lambda) - a)^2$ decrease?
2. Establish the validity of the foregoing procedure for the case where $u_{n+1} = au_n + v_n$, $u_0 = c$, $[J(u_n, v_n, \lambda) = \sum_{n=0}^{N} (u_n^2 + v_n^2) + \lambda(u_N - a)^2$.

2.21. Infinite Processes

In some cases the duration of the control process is of such an extent that it is a useful approximation to consider it to be unbounded. In place of an equation such as

$$f_N(c) = \min_{v} [h(c, v) + f_{N-1}(g(c, v))], \tag{2.21.1}$$

we obtain the equation

$$f(c) = \min_{v} [h(c, v) + f(g(c, v))], \tag{2.21.2}$$

where $f(c)$ is the return from the infinite process.

Two questions arise immediately:

(a) Does $f_N(c)$ possess a limit as $N \to \infty$ and does this limit function satisfy (2.21.2)?

(b) Does (2.21.2) possess a solution and is this solution unique?

Many interesting questions arise in this fashion, some of which we shall consider in Chapter 8.

2.22. Approximation in Policy Space

One approach to the solution of an equation such as (2.21.2) is to use the standard method of successive approximations,

$$f_n(c) = \min_v [h(c, v) + f_{n-1}(g(c, v))], \qquad n \geq 1,$$

$$f_0(c) = k(c). \tag{2.22.1}$$

Another approach is to approximate to the policy function $v(c)$. Let $v_0(c)$ be an initial policy and let an initial return function $f_0(c)$ be determined by $v_0(c)$, namely

$$f_0(c) = h(c, v_0) + f_0(g(c, v_0))$$

$$= h(c, v_0) + h(g(c, v_0), v_0^{(1)}) + \cdots, \tag{2.22.2}$$

where $v_0^{(1)} = v_0(g(c, v_0))$, and so on.

To obtain a new policy, let us choose $v_1(c)$ as the function which minimizes

$$h(c, v) + f_0(g(c, v)). \tag{2.22.3}$$

The new return function $f_1(c)$ is determined by the functional equation

$$f_1(c) = h(c, v_1) + f_1(g(c, v_1)), \tag{2.22.4}$$

and we continue in this fashion.

The problem of demonstrating that $f_1(c) \leq f_0(c)$ brings us in contact with the theory of positive operators. The function $f_1(c)$ satisfies the equality in (2.22.4) and $f_0(c)$ satisfies the inequality

$$f_0(c) = h(c, v_0) + f_0(g(c, v_0)) \geq h(c, v_1) + f_0(g(c, v_1)). \tag{2.22.5}$$

It is thus a matter of showing that

$$w(c) \geq w(g(c, v_1)) \tag{2.22.6}$$

implies that $w \geq 0$.

A number of interesting problems arise in this fashion which have surprising connections with many classical areas of analysis.

2.23. Minimum of Maximum Deviation

One of the advantages of the dynamic programming approach is that we can handle certain unconventional types of control processes in a rather direct fashion. Suppose, for example, that we are faced with the following control process:

(a) $u_{k+1} = g(u_k, v_k)$, $u_0 = c$,

(b) $J(\{u_k, v_k\}) = \max_{0 \le k \le N} [|u_k| + |v_k|]$.   (2.23.1)

To handle this problem, we use the basic functional property of the maximum operation, namely

$$\max [c_1, c_2, \ldots, c_N] = \max [c_1, \max [c_2, \ldots, c_N]].   (2.23.2)$$

Hence, we imbed the original criterion function in the family of criteria

$$J(\{u_k, v_k\}, b) = \max \left[b, \max_{0 \le k \le N} [|u_k| + |v_k|] \right].   (2.23.3)$$

Write

$$f_N(c, b) = \min_{\{u_k, v_k\}} J(\{u_k, v_k\}, b).   (2.23.4)$$

Then, the principle of optimality yields

$$f_N(c, b) = \min_v [f_{N-1}(g(c, v), \max [b, |c| + |v|])],   N \ge 1,   (2.23.5)$$

with

$$f_0(c, b) = \max [b, |c|].   (2.23.6)$$

The solution to the original problem is given by taking $b = 0$.

EXERCISES

1. Determine the structure of the optimal policy when $u_{k+1} = au_k + v_k$, $u_0 = c$.
2. How do we handle the problem when the criterion function has the form $\max_k |u_k| + \lambda \sum_{k=0}^N v_k$?

2.24. Discrete Stochastic Control Processes

The study of continuous stochastic control processes is not an easy one to present in a rigorous fashion. As we shall see in Chapter 4, continuous control processes of deterministic type require some more sophisticated considerations. Adding the stochastic features hardly diminishes the analytic level. (As we have noted in various prefaces, we will treat stochastic control processes in a separate volume.) Nonetheless, it is worth pointing out that many stochastic control processes of *discrete* type can be treated with little more than is required to treat the deterministic counterparts.

Consider, for example, the process defined by

(a) $u_{k+1} = g(u_k, v_k, r_k),$ $u_0 = c,$ (2.24.1)

(b) $J(\{v_k\}) = \exp_{r_k}\left[\sum_{k=0}^{N} h(u_k, v_k, r_k)\right].$

We assume here that the r_k are independent random variables with a specified distribution function $G(r)$.

The first interesting point that arises is that the control process is *not* completely defined by the defining equation and the criterion function. We must in addition specify the information that is available to the decision-maker at each point in time or, more generally, at each stage where a control variable is to be chosen.

There are two extremes:

(a)   The decisionmaker can examine the actual state of the system at each point where a decision is required.

(b)   The initial state, c, is known, but the system is not observable at any subsequent time.

In the first case, a policy function $v_k(c)$ represents the solution. Writing

$$f_N(c) = \min_{\{v_k(c)\}} J(\{v_k\}),   (2.24.2)$$

we readily obtain the functional equation

$$f_N(c) = \min_{v}\left[\int [h(c, v, r) + f_{N-1}(g(c, v, r))]\, dG(r)\right].   (2.24.3)$$

In the second case, a set of control values $v_0, v_1, \ldots, v_N$ represents the solution, and no corresponding functional equation exists. In the deterministic case, the two different conceptual approaches, event orientation versus time orientation, turned out to be equivalent. In the stochastic case, they are different both conceptually and analytically.

The difference lies in the information available to the decisionmaker at each stage. We see then that the theory of stochastic control processes is infinitely richer than the theory of deterministic control processes, since we can have all possibilities of partial information at each stage, ranging from none to total. Furthermore, we must now begin to take account of the type of observation of the system that is possible at each stage and the various costs involved.

Finally, we are forced to scrutinize most carefully the very concept of "uncertainty." This opens up even more intriguing types of investigations.

Miscellaneous Exercises

1. Write $f_p(c_1, c_2, \ldots, c_N) = (\sum_{i=1}^{N} c_i^p)^{1/p}$, $p \geq 0$. Show that $f_p(c_1, c_2, \ldots, c_N)$
   $= f_p(f_p(c_1, c_2, \ldots, c_{N-1}), c_N)$.

2. Let the c_i be positive. What is the limit of $f_p(c_1, c_2, \ldots, c_N)$ as $p \to \infty$?
   What is the limit, if any, as $p \to -\infty$?

3. Let $f_N(c) = \max a_1 a_2 \cdots a_N$ over the region $\sum_{i=1}^{N} a_i = c$, $a_i \geq 0$. Show that
   $f_N(c) = \max_{0 \leq a_N \leq c} [a_N f_{N-1}(c - a_N)]$. Use the fact that $f_N(c) = k_N c^N$
   where k_N is independent of c to deduce the value of k_N, and thus
   the arithmetic-geometric mean inequality for nonnegative quantities,
   $a_1 a_2 \cdots a_N \leq ((a_1 + a_2 + \cdots + a_N)/N)^N$ with equality only if $a_1 = a_2 = \cdots = a_N$.

4. What relations holds between f, g, and h if $\max_{x \geq 0} (f(x) - g(x)y) = h(y)$? (See

   A. F. Timan, *Trans. Acad. Sci. USSR*, Math. series—IAN, **29**, 1964, pp. 35–47.)

5. The functional equation $f(c) = \min_v [c^2 + v^2 + f(ac + v)]$ possesses a
   solution of the form $f(c) = kc^2$ for an appropriate choice of k. Does it
   possess any other solutions which are convex and zero at the origin? See
   R. Bellman, "Functional Equations in the Theory of Dynamic Programming—XVI: An Equation Involving Iterates," *JMAA*, 1970, to
   appear.

6. Consider the maximization of $\sum_{i=1}^{N} g_i(x_i)$ over the region $\sum_{i=1}^{N} x_i = c$,
   $x_i \geq 0$. Write $f_N(c) = \max [\sum_i g_i(x_i)]$ and show that $f_N(c) = \max_{0 \leq x_N \leq c}$
   $[g_N(x_N) + f_{N-1}(c - x_N)]$, $N \geq 2$. Determine $f_N(c)$ and the optimal policy
   in the cases where $g_i(x) = c_i x^2$, $g_i(x) = c_i x^{1/2}$, $c_i > 0$. (In the general
   case there are a number of interesting questions connected with this
   problem arising in the study of the optimal allocation of resources. See

   T. Furubayashi, *Math. Rev.*, no. 7202, **34**, 1967.
   W. Karush, "A Theorem in Convex Programming," *Naval Res. Logs.
   Quart.*, **6**, 1959, pp. 245–260.
   W. Karush, *Manag. Sci.*, **9**, 1962, pp. 50–72.)

7. Consider the foregoing maximization problem using a Lagrange multi-
   plier, $f_N(c, \lambda) = \max_{x_i \geq 0} [\sum_{i=1}^{N} g_i(x_i) - \lambda \sum_{i=1}^{N} x_i]$. What are the advantages and disadvantages of this approach?

8. What connection is there, if any, between the "price" λ and the "price"
   $f_N'(c)$?

9. How could one treat the problem of maximizing $\sum_{i=1}^{N} g_i(x_i)$ subject to
   the constraints $\sum_{i=1}^{N} a_i x_i \leq c_1$, $\sum_{i=1}^{N} b_i x_i \leq c_2$, $x_i \geq 0$?

10.   Consider the case where $g_i(x_i) = b_i x_i$, $b_i > 0$, and the x_i are constrained to be positive integers on zero. (The maximum function is the so-called "knapsack function." There are some interesting connections with group theory. See

P. C. Gilmore and R. E. Gomory, "A Linear Programming Approach to the Cutting Stock Problem—I" *Oper. Res.*, **9**, 1961, pp. 849–859.
P. C. Gilmore and R. E. Gomory, "A Linear Programming Approach to the Cutting Stock Problem—II," *Oper. Res.*, **11**, 1963, pp. 863–888.
P. C. Gilmore and R. E. Gomory, "The Theory and Computation of Knapsack Functions," *Oper. Res.*, **14**, 1966, pp. 1045–1074.
H. Greenberg, "An Algorithm for the Computation of Knapsack Functions," *J. Math. Anal. Appl.*, **26**, 1969, pp.
J. F. Shapiro and H. M. Wagner, "A finite Renewal Algorithm for the Knapsack and Turnpike Models," *Oper. Res.*, **15**, 1967, pp. 319–341.)

11.   Discuss the maximization of $R_N(x, y) = \sum_{i=1}^{N} (gy_n^p + h(x_n - y_n)^p)$, where $x_{n+1} = ay_n + b(x_n - y_n)$, $x_0 = c$, $a, b, g, h \geq 0$.

(J. R. Bagwell, C. S. Beightler, and J. P. Stark, "On a Class of Convex Allocation Problems," *J. Math. Anal. Appl.*, **25**, 1969.)

12.   If A is a positive definite matrix, then $\min_x [(x, Ax) - 2(x, y)] = -(y, A^{-1}y)$.

13.   Consider the system of equations $Ax - By = u$, $Ay + Bx = v$. If A is negative definite, these can be considered to be the variational equations arising from determining the maximum over y and the minimum over x of $(y, Ay) - (x, Ax) + 2(x, By) + 2(u, x) - 2(v, y)$. Does it make any difference in which order we carry out these operations?

14.   Consider the equation $Ay = u$, where A^{-1} exists, but A is not necessarily symmetric. Consider the problem of maximizing the function $(x, Bx) + (y, By) + 2(x, Ay) - 2(u, x) - 2(v, y)$, over x and y where B is negative definite. Show that the variational equations are $Bx + Ay = u$, $A^T x + By = v$. As $B \to 0$, does the solution of this system approach the solution of $Ay = u$? (This is a "regularization" procedure in the sense of Tychonov. See

R. Bellman, R, Kalaba, and J. Lockett, *Numerical Inversion of the Laplace Transform*, Elsevier, New York, 1966.
R. Bellman and R. S. Lehman, "Functional Equations in the Theory of Dynamic Programming—IX: Variational Analysis, Analytic Continuation, and Imbedding of Operators," *Proc. Nat. Acad. Sci.*, *USA*, **44**, 1958, pp. 905–907.
R. Lattes and J. L. Lions, *The Method of Quasi-reversibility: Applications to Partial Differential Equations*, Elsevier, New York, 1969.)

15. Consider the problem of minimizing

$$g(a_1) + \frac{g(a_2)}{a_1} + \frac{g(a_3)}{a_1 a_2} + \cdots + \frac{g(a_M)}{a_1 a_2 \cdots a_{M-1}}$$

over all a_i subject to $a_1 a_2 \cdots a_M = c$, $a_i \geq 1$. (See

R. Bellman, "A Dynamic Programming Solution to a Problem in Heavy Water Production," *Nuclear Sci. Eng.*, **2**, 1957, pp. 523–525, for the background of this problem.)

16. Consider the problem of minimizing $Q(x) = \sum_{k=1}^{N} \left(\sum_{v \mid a_k} x_v^2 \right)$, over the x_i where $x_1 = 1$. Here $v \mid a_k$ signifies that v divides the integer a_k. (See

R. Bellman, "Dynamic Programming and the Quadratic Form of Selberg," *J. Math. Anal. Appl.*, **15**, 1966, pp. 30–32.)

17. Let a system S be described by a state vector p and suppose that the objective is to transform p into a state which is immediately recognizable Let $T(p, q)$ be a family of transformations dependent on the control variable q with the property that $p_1 = T(p, q)$ is dependent on the control variable q with the property that $p_1 = T(p, q)$ is a state of S for all p and all admissible q. Let $f(p) =$ the minimum number of transformations required to recognize p. Then $f(p) = 1 + \min_q f(T(p, q))$. (See

R. Bellman, "Dynamic Programming, Pattern Recognition, and Location of Faults in Complex Systems," *J. Appl. Prob.*, **3**, 1966, pp. 268–271.
    For a study of functional equations of this nature see

R. Bellman, *Dynamic Programming*, Princeton Univ. Press, Princeton, New Jersey, 1957.
M. Kwapisz, "On a Certain Functional Equation." *Colloq. Math.*, **18**, 1967, pp. 169–179.)

18. Consider the problem of determining the maximum of $L_n(v) = v_1 + v_2 + \cdots + v_n$ over all v_i subject to the constraints $r_i \geq v_i \geq 0$, $v_1 \leq b_1$, $v_1 + v_2 \leq b_2, \ldots, v_1 + v_2 + \cdots + v_k \leq b_k$, $v_2 + v_3 + \cdots + v_{k+1} \leq b_{k+1}, \ldots, v_{n-k+1} + \cdots + v_n \leq b_n$. (See

R. Bellman, "On a Dynamic Programming Approach to the Caterer Problem—I," *Manag. Sci.*, **3**, 1957, pp. 270–278.)

19. Consider the problem of minimizing $\sum_{i=1}^{N} c_i x_i$ subject to $\sum_{i=1}^{k} x_i \geq r_k$, $k = 1, 2, \ldots, N$, $x_i \geq 0$. (See

R. Bellman, "Dynamic Programming and the Smoothing Problem," *Manag. Sci.*, **3**, 1956, pp. 111–113).

20. Solve $f(x) = \max_{0 \le y \le x} [a(y) + (1 - a(y))f(x - y)]$ under suitable hypotheses on $a(y)$.

21. Solve $f(x) = \min [1 + xf(1), 1 + f(ax)], 0 < x \le 1, f(0) = 0, 0 < a < 1$.

22. Solve $f(x, y) = \max [p_1(r_1x + f((1 - r_1)x, y)), p_2(s_1y + f(x, (1 - s_1)y))]$ assuming that $x, y \ge 0, 0 < r_1, s_1 < 1, p_1, p_2 \ge 0$. (For the foregoing three problems, see

R. Bellman, *An Introduction to the Theory of Dynamic Programming*, The RAND Corp. R-245, 1953.)

23. Let $0 < a \le t \le b$ and let s be determined to minimize $\max_t |s - t|$. Show that $s_{min} = (a + b)/2$ and $\min_s \max_t |s - t| = (b - a)/2$.

24. Under the same conditions show that

$$\min_s \max_t \{|s - t|/t\} = (b - a)/(b + a),$$

and $s_{min} = 2ab/(a + b)$. (See

G. Polya, "On the Harmonic Mean of Two Numbers," *Amer. Math. Monthly*, **LVII**, 1950, pp. 26–28.)

25. What connections are there between the following problems:
(1) $\min_v \sum_{n=0}^{N} (u_n^2 + v_n^2)$, subject to $u_{n+1} = au_n + v_n, u_0 = c$, and $|u_n| \le k, n = 0, 1, \ldots, N$; and
(2) $\min_v [\sum_{n=0}^{N} (u_n^2 + v_n^2) + \lambda \max_{0 \le n \le N} |u_n|]$, subject to $u_{n+1} = au_n + v_n, u_0 = c$?
Obtain a functional equation for the solution of the second problem.

26. Consider the problem of minimizing $\sum_{n=0}^{N} (u_n^2 + v_n^2)$ with respect to v where $u_{n+1} = au_n + v_n, u_0 = c$. Suppose that we suboptimize, setting $v_n = bu_n, n = 0, \ldots, N$. How does one determine b?

27. Consider the vector case where we wish to minimize

$$\sum_{n=0}^{N} [(x_n, x_n) + (y_n, y_n)]$$

with respect to y where $x_{n+1} = Ax_n + y_n, x_0 = c$. Suppose that we suboptimize, setting $y_n = Bx_n, n = 0, \ldots, N$, where B is a $k \times N$ matrix

$$B = \begin{pmatrix} b_{11} & b_{12} & \cdots & b_{1k} \\ b_{21} & b_{22} & \cdots & b_{2k} \\ \vdots & & & \\ b_{N1} & b_{N2} & \cdots & b_{Nk} \end{pmatrix}$$

How does one determine the b_{ij}?

28. Given the linear difference relation $x_{n+1} = Ax_n + y_n, x_0 = c$, can one always determine $y_0, y_1, \ldots, y_{N-1}$ so that $x_N = d$?

29. Can it always be done if we introduce local constraints, $\|y_n\| \le k$? Global constraints, $\sum_{n=0}^{N} (y_n, y_n) \le k$?

30. What is the connection between the foregoing problem and minimizing $\sum_{n=0}^{N} (y_n, y_n) + \lambda \|x_N - d\|$? (This is related to the concept of "controllability". See

J. N. Johnson, "Controllability in Nonlinear Control Systems," Dept. of Mathematics, Univ. of California, June 1969.
R. E. Kalman, Y. C. Ho and K. S. Narenda, "Controllability of Linear Dynamical Systems," *Contrib. Differ. Equations*, **1**, 1963, pp. 189–213).

BIBLIOGRAPHY AND COMMENTS

2.1. There are a number of books on dynamic programming now available. Let us refer to

R. Aris, *Discrete Dynamic Programming*, Blaisdell, New York, 1964.
R. Bellman, *Dynamic Programming*, Princeton, Univ. Press, Princeton, New Jersey, 1957.
R. Bellman, *Adaptive Control Processes: A Guided Tour*, Princeton Univ. Press, Princeton, New Jersey, 1961.
R. Bellman and S. Dreyfus, *Applied Dynamic Programming*, Princeton Univ. Press, Princeton, New Jersey, 1962.
S. Dreyfus, *Dynamic Programming and the Calculus of Variations*, Academic Press, New York, 1965.
A. Kaufmann, *Graphs, Dynamic Programming, and Finite Games*, Academic Press, New York, 1967.
A. Kaufmann and R. Cruon, *Dynamic Programming: Sequential Scientific Management*, Academic Press, New York, 1967.

For a discussion of the genesis of dynamic programming, see

R. Bellman, "Dynamic Programming: A Reluctant Theory," *New Methods of Thought and Procedure*, (F. Zwicky and A. Wilson, eds.), Springer, Berlin, 1967.

2.4. A number of papers have been devoted to the structure of dynamic programming processes. See

R. F. Arnold and D. L. Richards, "Monotone Reduction Algorithms," in *Recent Mathematical Advances in Operations Research*, No. 129, Univ. of Michigan, Ann Arbor, Summer 1964.
U. Bertele and F. Brioschi, *Paramentrization in Nonsocial Dynamic Programming*, 1969, to appear.
T. A. Brown and R. E. Strauch, "Dynamic Programming in Multiplicative Lattices," *JMAA*, **12**, 1965, pp. 365–370.
M. F. Clement, "Modéle Categorique de processus de decisions, théorème de Bellman, Applications," *C. R. Acad. Sci. Paris*, **268**, 1969, pp. 1–4.
D. O. Ellis, *An Abstract Setting for the Notion of Dynamic Programming*, The RAND Corp., P-783, 1955.
S. E. Elmaghraby, "The Concept of 'State' in Discrete Dynamic Programming," *JMAA*, to appear.
I. Fleischer and A. Kooharian, "Optimization and Recursion," *J. Soc. Indus. Appl. Math.*, **12**, 1964, pp. 186–188.
R. Fortet, "Properties of Transition Mappings in Dynamic Programming," (in French), *METRA*, **2**, 1963, pp. 79–97.

H. Halkin, "The Principle of Optimal Evolution," in *Nonlinear Differential Equations and Nonlinear Mechanics*, Academic Press, New York, 1963, pp. 284–302.

P. Ivanescu, "Dynamic Programming with Bivalent Variables," no. 7261 *Mat. Vesnik*, **3(18)**, 1966, pp. 87–99; see *Math. Rev.*, no. 7261, **34**, 1967.

S. Karlin, "The Structure of Dynamic Programming Models," *Naval Res. Logs. Quart.*, **2**, 1955, pp. 285–294.

L. G. Mitten, "Composition Principles for Synthesis of Optimal Multistage Processes," *Oper. Res.*, **12**, 1964, pp. 610–619.

P. J. Wong, "A New Decomposition Procedure for Dynamic Programming," **18**, 1970, pp. 119–131.

A Wouk, "Approximation and Allocation," *J. Math. Anal. Appl.* **8**, 1964, pp. 135–143.

2.5.   Recurrence relations of this nature can be used to treat a number of combinatorial and scheduling problems. See

R. Bellman, "An Application of Dynamic Programming to the Coloring of Maps," *ICC Bull.*, **4**, 1965, pp. 3–6.

See, for a discussion of analogous problems and particularly for a discussion of a number of the classic puzzles,

R. Bellman, K. L. Cooke, and J. Lockett, *Algorithms, Graphs, and Computers*, Academic Press, New York, 1970.

A number of other references will be found there.

2.7.   A point worth stressing is that frequently a solution of a functional equation arising in dynamic programming is most easily found and expressed in terms of the structure of the optimal policy.

2.13.   See

I. Fleischer and H. Young, "The Differential Form of the Recursive Equation for Partial Optima," *J. Math. Anal. Appl.* **9**, 1964, pp. 294–302.

2.14.   The reduction of the solution of variational problems involving linear equations and quadratic criteria to matrix iteration in the discrete case, an extension of continued fractions, and to the Riccati equation in the continuous case was first given in

R. Bellman, "On a Class of Variational Problems," *Quart. Appl. Math.*, **14**, 1957, pp. 353–359.

See the first volume in this series,

R. Bellman, *Introduction to the Mathematical Theory of Control Processes*, Vol. I: *Linear Equations and Quadratic Criteria*, Academic Press, New York, 1968.

for extensive discussion and many further references.

2.15.   We shall discuss "inverse problems" again in a later chapter. See

R. Bellman, "Dynamic Programming and Inverse Optimal Problems in Mathematical Economics," *J. Math. Anal. Appl.*, to appear.

R. Bellman and R. Kalaba, "An Inverse Problem in Dynamic Programming and Automatic Control, *J. Math. Anal. Appl.*, **7**, 1963, pp. 322–325.

R. Bellman, H. Kagiwada, and R. Kalaba, "Dynamic Programming and an Inverse Problem in Neutron Transport Theory," *Computing*, **2**, 1967, pp. 5–16.

R. Bellman and J. M. Richardson, "A Note on an Inverse Problem in Mathematical Physics," *Quart. Appl. Math.*, **19**, 1961, pp. 269–271.

2.16. This example is taken from the monograph

R. Bellman, I. Glicksberg, and O. Gross, *Some Aspects of the Mathematical Theory of Control Processes*, The RAND Corp. R-313, 1958.

See also

R. Bellman, "A Variational Problem with Constraints in Dynamic Programming," *J. Soc. Indus. Appl. Math.*, **4**, 1956, pp. 48–61.

2.18. See

H. Everett, III, "Generalized Lagrange Multiplion Method for Solving Problems of Optimal Allocation of Resources," *Oper. Res.* **11**, 1963, pp. 399–417.

2.20. See

T. Butler and A. V. Martin, "On a Method of Courant for Minimizing Functionals," *J. Math. & Phys.*, **41**, 1962, pp. 291–299.

2.21. For a brief exposition of functional equations, see

R. Bellman, "Fuhctional Equations," *Handbook of Mathematical Psychology*, Vol. III, Wiley, New York, 1965.

For a detailed discussion of classical functional equations, see the book

J. Aczel, *Lectures on Functional Equations and Their Applications*, Academic Press, New York, 1966.

Many interesting and difficult problems arise in the consideration of functional equations arising from unbounded processes. See, for example,

R. Bellman, "Some Functional Equations in the Theory of Dynamic Programming—I: Functions of Points and Point Transformations," *Trans. Amer. Math. Soc.*, **80**, 1955, pp. 51–71.

R. Bellman and T. Brown, "Projective Metrics in Dynamic Programming," *Bull. Amer. Math. Soc.* **71**, 1965, pp. 773–775.

E. B. Keeler, "Projective Metrics and Economic Growth Models," *RM-6153-PR*, September 1969.

2.22. The method of "approximation in policy space" brings us into contact with the theory of quasilinearization. See

R. Bellman and R. Kalaba, *Quasilinearization and Nonlinear Boundary-value Problems* Elsevier, New York, 1965.

E. F. Beckenbach and R. Bellman, *Inequalities*, Springer, Berlin 1961.

2.23. Problems of determining the minimum of maximum deviation arise in a number of important control processes. See, for example,

M. Ash, *Optimal Shutdown Control of Nuclear Reactors*, Academic Press, New York, 1966.

R. Bellman, M. Ash, and R. Kalaba, "On Control of Reactor Shutdown Involving Minimal Xenon Poisoning," *Nuclear Sci. Eng.*, **6**, 1959, pp. 152–156.

2.24. We shall discuss discrete stochastic control processes in detail in Volume III. See the books cited in Section 2.1. and

R. Howard, *Dynamic Programming and Markov Processes*, Wiley, New York, 1960.

H. Mine and S. Osaki, *Markovian Decision Problems*, Elsevier, New York, 1970.

3

––––

COMPUTATIONAL ASPECTS
OF DYNAMIC PROGRAMMING

3.1. Introduction

Let us now devote some time to an examination of the computational feasibility of a numerical solution of discrete control processes based on the functional equations provided by dynamic programming. A typical optimization problem is that of minimizing the scalar function

$$J(\{x_n, y_n\}) = \sum_{n=0}^{N} h(x_n, y_n), \qquad (3.1.1)$$

with respect to the y_n where the vectors x_n and y_n are subject to the relations

$$x_{n+1} = g(x_n, y_n), \qquad x_0 = c, \qquad n = 0, 1, \ldots, N-1. \qquad (3.1.2)$$

This has been converted, as we have seen in Chapter 2, into the problem of determining a sequence of functions $\{f_k(c)\}$, $k = 0, 1, \ldots, N$, by means of the recurrence relation

$$f_{k+1}(c) = \min_y [h(c, y) + f_k(g(c, y))],$$

$$f_0(c) = \min_y h(c, y), \qquad (3.1.3)$$

where c belongs to a specified region. Additional constraints on the y_n of local nature will be reflected by additional constraints on the vector y in (3.1.3) as we have seen. As we shall again point out, below, these local constraints considerably simplify the numerical solution.

In this chapter we wish to discuss some aspects of the numerical solution of (3.1.3) using contemporary digital computers, against a background of the application of reasonably straightforward algorithms. In subsequent chapters we shall describe various devices and strategems which can be used in favorable situations to overcome the considerable obstacles to a direct approach in the case where the dimension of the vector c is large. Some additional methods of importance are cited at the end of the chapter.

No *uniform* method now exists for obtaining the numerical solution of control processes, nor do we ever expect to find one. We might well add "fortunately," since this area is a happy-hunting ground for mathematicians and since considerable ingenuity is stimulated, and skill generated, by the constant search for effective and improved methods.

3.2. Discretization

Our aim is to obtain a numerical determination of $f_N(c)$ for some fixed value of N and a particular value of c using only the capabilities of a digital computer. To accomplish this we must, in some fashion, discretize all functions and arithmeticize all operations.

We suppose that the components of c are each restricted to lie in some finite interval which, by means of preliminary normalization, we may conveniently take to be $[0, 1]$. In this interval we restrict our attention to the fixed set of grid points $\{k\Delta\}$, where $k = 0, 1, \ldots, M$, with $M\Delta = 1$. Here Δ is an assigned value, the grid size, which then determines $M + 1$, the number of grid points. A function $f(c)$ originally defined over the unit R-dimensional cube is now taken to be defined solely by its values at the uniform network of points

$$[k_1\Delta, k_2\Delta, \ldots, k_R\Delta], \tag{3.2.1}$$

$0 \le k_i \le M, i = 1, 2, \ldots, R$, where R is the dimension of the state vector x. From this unsophisticated point of view, a function $f(c)$ is merely a heterogeneous collection of $(M + 1)^R$ numbers.

The values of the components of $g(c, y)$ will not generally, as might be expected, coincide with those of a grid point, as given in (3.2.1), when c is a grid point and y is an admissible control variable. Hence, in order to evaluate the quantity $f_k(g(c, y))$ in terms of the values of $f_k(c)$ at the grid points, the only values accessible to us, some type of interpolation procedure will be required. One method is to use the functional value at the nearest lattice point; another is to apply linear interpolation to the values at the surrounding lattice points. In some cases it may be profitable to use a more sophisticated interpolation procedure.

It can also occur that one or more of the components of $g(c, y)$ lie outside of the interval $[0, 1]$. As before, we can resolve this difficulty by using the nearest lattice point. Thus, for example, if one component of $g(c, y)$ lies outside the interval, we can define

$$g_1(c, y) = 0, \quad \text{for } g_1(c, y) \le 0,$$
$$= 1, \quad \text{for } g_1(c, y) \ge 1. \tag{3.2.2}$$

Here $g_1(c, y)$ denotes the first component of $g(c, y)$.

Alternatively, we can avoid the problem by admitting as values of the control vector y only those which maintain the desired bounds,

$$0 \le g_i(c, y) \le 1, \tag{3.2.3}$$

$i = 1, 2, \ldots, R$. We shall pursue this idea in Chapter 10.

Occasionally, as illustrated in the Miscellaneous Exercises, we can use some ingenuity to ensure that the original domain is maintained from stage to stage.

An "expanding grid," the foregoing phenomenon, is frequently encountered in the theory of partial differential equations. A much larger set of initial data may be required to determine a particular set of values at some subsequent time. Sometimes there is a valid physical reason for this; sometimes it merely represents a defect in our analytical-computational tools. See Figure 3.1.

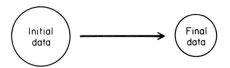

Figure   3.1

In each particular case a stability investigation is necessary to determine whether approximations and constraints of the type imposed in (3.2.2) and (3.2.3) produce any significant change in either the return function or the optimal policy. If they do, the computational approach must be revised. Frequently, however, this instability yields a significant clue which can be utilized in ascertaining the structure of the optimal policy and in determining an accurate estimate of the solution.

3.3. Minimization

Let us now turn to the minimization procedure required for the use of (3.1.3). In general, the values of y will be restricted to another finite set since we are momentarily pledged to the use of arithmetic techniques. Again, with a suitable normalization, we can suppose that each component of y is restricted to the interval $[0, 1]$, assuming only the values

$$0, \delta, 2\delta, \ldots, K\delta, \qquad K\delta = 1. \qquad (3.3.1)$$

The step δ will almost invariably be different from the previous grid variable Δ.

The minimum in (3.1.3) with respect to y is then determined by a direct search over $(K + 1)^R$ values, taking the dimension of y at the moment to be the same as the dimension of x. In many important cases, however, the dimension of y will be significantly less than that of x, which is an important point as far as time for the minimization operation is concerned. Furthermore, as we indicate below, effective calculation in many cases requires a search which is a bit more than direct.

The calculation begins with the determination of the function

$$f_0(c) = \min_{y} h(c, y), \qquad (3.3.2)$$

along the lines described above for each particular value of c. The set of values of $f_0(c)$ at the assigned grid points is then stored in rapid-access storage preliminary to a determination of $f_1(c)$ at these same grid points using (3.1.3). Simultaneously, the policy function $y(c, 0)$, i.e., the function $y(c)$ yielding the minimum in (3.3.2) is determined. This function, of course, as previously mentioned, need not be single-valued. There is no difficulty in storing the various values which yield the minimum, when this situation arises.

Once $f_1(c)$ and $y(c, 1)$ have been determined and stored in the rapid-access storage, there is no longer any need for either of the functions $f_0(c)$ or $y(c, 0)$. Both are printed out, if their values are desired subsequently, and then discarded in order to provide as much space as possible in the rapid-access storage. The process, which is iterative, now continues with the determination of $f_2(c)$, $y(c, 2)$, $f_3(c)$, $y(c, 3)$, and so forth.

Frequently, we can use a knowledge of $y(c, 0)$ to guide the search for $y(c, 1)$; see Section 3.20.

3.4. Storage Requirements: Discretion in the Use of Discretization

As we have already indicated, the foregoing algorithm requires that the functions $g(c, y)$, $h(c, y)$, and $f_k(c)$ be available in order to determine $f_{k+1}(c)$. In the majority of cases of importance, the functions $g(c, y)$ and $h(c, y)$ can be easily calculated as we need them using some explicit analytic expression.

Hence, the burden of the computational process falls upon the storage and retrieval of the function $f_k(c)$. If we follow the direct method described above, we require the capacity to store and retrieve $(M + 1)^R$ values. Let us see what this means.

At the moment, we can regard with equanimity a requirement for 10^6 ten or twenty digit numbers in rapid-access storage, with about 10^8 in slow-access storage. Over the next ten or twenty years, we can expect these figures to increase to about 10^8 and 10^{10}, respectively. As we shall see in a moment, the precise values are not of significance, only an order of magnitude. In other words, whether these figures become 10^{10} and 10^{12}, respectively, is not significant for the argument here. A difference, however, between 10^8 and 10^9 can be crucial for the success of the more sophisticated procedures we shall subsequently describe. By "rapid-access," we mean here a time of operation of about 10^{-7} seconds; by "slow-access," we mean a time of about 10^{-5} seconds. The same comments apply here as a moment ago.

Taking $M + 1 = 100$, we see that processes involving state vectors of dimension one or two introduce no difficulty. A state vector of dimension three, however, produces a requirement which strains contemporary abilities. While storing $f_k(c)$, we must also provide some room for $f_{k+1}(c)$ and $y(c, k + 1)$, and for the details of the program; hence, a more appropriate bound is $3(M + 1)^R$. A process with $R = 4$ has an associated requirement of

$$(100)^4 = 10^8; \tag{3.4.1}$$

$R = 5$, with its requirement of 10^{10}, appears to escape the projected capacity of even ten years in the future—speaking always in terms of the rudimentary procedures described above and assuming no major breakthrough.

It is clear from these simple calculations that even if our estimates of computer capacity are pessimistic by a factor of one hundred, or one thousand (or one million), we cannot expect to treat high-dimensional control processes, e.g., $R = 10$ or 20, using the methods presented above, in a routine fashion at any time in the near future.

This is a pleasing thought to the mathematician. It means that mathematical sophistication and ingenuity must be employed to break the spell of this "curse of dimensionality" that now holds us in thrall. Only an adroit exploitation of the structure of a particular process, combined with the power of the computer, will enable us to handle large systems.

It is possible, but not probable, that an unforeseen breakthrough will allow us to revise considerably some of the foregoing estimates in an upwards direction. This would nevertheless make little difference to the mathematician since there are a plethora of control processes of dimension 100 or 1000 or higher which clamor for treatment.

3.5.   Time Requirements

The determination of the function $f_N(c)$, as outlined above, is an iterative operation. By this we mean that $f_1(c)$ is determined by $f_0(c), f_2(c)$ by $f_1(c)$, and generally

$$f_N(c) = T(f_{N-1}(c)), \tag{3.5.1}$$

where the transformation T may or may not be independent of N. The time, therefore, required to calculate $f_N(c)$ is essentially N times the time required to calculate $f_1(c)$. This estimate may vary a bit if T depends in some essential way on N, but the order of magnitude is correct. This basically *linear* dependence of computing time on N is a remarkable feature of dynamic programming, and it stands in sharp contrast to the situation presented by other methods where computing time may be proportional to N^2, N^3, or higher.

The choice of Δ determined the rapid-access storage requirements examined above. The choice of both Δ and δ determines the time required to carry out all of the searches required in the calculation of the optimal policy $y(c, k)$, and thus fixes the time per stage.

As usual in numerical work, we are torn between two strong desires. On one hand, we want very much to make both Δ and δ small in order to ensure accuracy; on the other hand, we wish to avoid making these quantities too small since computing time, following the procedures described above, is directly proportional to $1/\Delta$ and $1/\delta$, or to powers of these quantities, depending on the dimension.

It must, and should, be admitted that there are no completely reliable a priori ways of determining appropriate values of Δ and δ. As in the numerical solution of ordinary and partial differential equations and moreover as in all areas of numerical analysis, experience, physical intuition, and trial and error play almost equally important parts in the final choice of parameters. As we shall see below, we can establish some a priori bounds on the error due to choices of Δ and δ, but these are usually much too pessimistic.

In many cases, fortunately, time is not the basic constraint. This will become more and more the case as computers become faster and a greater degree of parallel processing is introduced. By the term " parallel processing " we mean the ability of the computer to use separate components for different tasks at the same time. The fundamental and governing limitation in the application of dynamic programming will continue to be the availability of rapid-access storage. This is the crucial factor which determines whether a problem can be tackled at all.

Hence, our principal objective will most often be to provide analytical techniques for trading *time*, which is expensive but unbounded, for *rapid-access storage*, which is cheap but limited. This type of trade-off introduces many new types of mathematical questions, some of which we shall discuss below.

3.6. Discussion

We have in the previous chapters gone into some detail concerning the various stages of problem-solving. In the foregoing section, we have discussed some aspects of feasibility of numerical solution. One point we wish to stress is that "feasibility" is a thoroughly time-dependent concept. As mathematical expertise grows, as new methods are developed, as computers grow in power, as programmers become more adroit in the use of computers, certain types of approaches gradually become "feasible." At the present time, many classes of problems require special approaches and complex programs. Over time, we know that these problems will gradually fit into general approaches, thereby enabling ad hoc methods to be discarded.

One of the reasons, however, for our constant harping on the theme of "numerical solution of numerical questions" is that we observe from the history of mathematics that new methods developed for particular problems, methods to which we have just applied the pejorative term "ad hoc," can easily blossom into new theories which can be applied in surprising fashions to both new and old problems. Consequently, there is strong motivation for challenging the ingenuities and skills of mathematicians in this fashion.

We have not discussed here another type of feasibility problem usually ignored in mathematical texts, namely that of developing carefully pruned versions of sophisticated methods which can readily be taught to, and used by, engineers, economists, biologists and others, who are primarily interested in specific responses to specific questions and not in the enveloping mathematical theory. There are numerous pedagogical problems involved in introducing new methods into new areas in conjunction with the use of computers. All of this, part of a systems approach to education is a sadly neglected area.

3.7. Simplified Policies

As an example of the type of simplification that can reduce computing time by a sizeable factor, consider the problem of determining a path of minimum time between two points P and Q in the plane; see Figure 3.2. In the dynamic programming approach, we are required to determine a direction $v = v(R)$ at each point R lying in some allowable region containing P and Q. We discretize

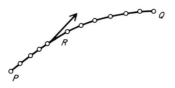

Figure   3.2

this minimization problem in an obvious way by allowing at each point R a choice of only a finite number of directions, say 10 or 20. It is clear, however, that if the time interval between decisions is made sufficiently short, we can reduce the number of directions to three; see Figure 3.3.

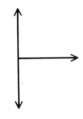

Figure   3.3

Furthermore, if we knew that the curve was an ascending one, or a descending one, we could reduce this choice still further to one of two allowable directions.

We suspect that an adequate approximation can be obtained in this way, assuming a sufficiently short time between decisions, since we know that a continuous curve can be arbitrarily approximated by a "staircase" curve, as indicated in Figure 3.4.

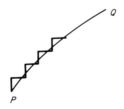

Figure   3.4

A new type of approximation problem thus appears in connection with numerical techniques of the foregoing nature, "approximation in policy space" versus the conventional approximation in function space.

Approximations, particularly plausible ones, must always be examined with a critical eye. Consider, for example, the classic example of two curves arbitrarily close in the sense of the usual Euclidean metric with the length of one *twice* the length of the other; see Figure 3.5. To obtain this "approximation," we begin by bisecting the segment PQ. On the segments PR and RQ we construct equilateral triangles. We now repeat this process on the line

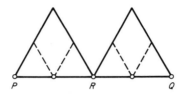

Figure   3.5

segments PR and RQ, as indicated by the dotted lines, and then again, and so on. In this way we can, by stopping at a suitable stage, obtain a polygonal curve arbitrarily close to the line segment PQ with length precisely twice that of PQ.

Suppose then we are attempting to determine a path of shortest time from P to Q with the restriction that we could pursue only the two directions shown in Figure 3.6. We would always obtain, then, no matter how the uniform step

Figure   3.6

size was decreased, a path which took twice as long as the optimal path, the straight line joining P and Q, assuming a uniform velocity.

It follows that the discretization of policy space must always be done with care. In particular, it is essential to investigate the convergence of the discrete process as both Δ and δ tend to zero. We will pursue some simple aspects of this interesting type of question in the following section. In Chapter 7 we consider some questions associated with the effect of letting the step size in time tend to zero.

What is important to note is that the existence of constraints on the choice of control variables considerably simplifies the computational algorithm we have been examining. Each constraint eliminates a class of admissible policies. There is thus considerable motivation in employing as realistic a description of a physical process as possible when dynamic programming is utilized. The more realistic the description, the narrower the choice of possible control variables.

3.8. Stability

Three types of errors in evaluation can be committed in the calculation of the sequence $\{f_k(c)\}$ following the procedures outlined above:

(a) errors in the evaluation of $h(c, y)$ and $g(c, y)$;
(b) errors in the evaluation of $f_k(g(c, y))$; and
(c) error in the determination of the minimum over y.

It follows that we must face the question of computational stability. Does a small error committed at a particular stage of the calculation become uncomfortably amplified over time? We would suspect that this is not the case, from the very concept of dynamic programming and the use of the principle of optimality. A proof that the computational process outlined above is stable, with a reasonable definition of the term, turns out to be quite easy. Note that stability is a relative concept. We are always examining the persistence of a specified property relative to a particular type of transformation.

Errors in the evaluation of $h(c, y)$ and $g(c, y)$ can arise in two ways. In the first place, we may have to replace a function such as e^t by some rational approximation which can be evaluated by a digital computer. In the second place, even if all the functions appearing as components of $h(c, y)$ and $g(c, y)$ are rational functions to begin with, their evaluation will be exact only to ten or perhaps twenty significant figures. At some point, a round-off error must be committed.

3.9. Computing as a Control Process

The problem that confronts us is that of choosing an algorithm and organizing its components in such a way as to minimize the errors in the end results of the calculation. This is a multistage process with a terminal state criterion and, usually, constraints on the course of the process, e.g., time and rapid-access storage.

Although it is clearly a deterministic control process in origin, we usually convert it into a stochastic control process for the sake of simplicity. Rather than attempt to deal directly with round-off errors as functions of the state and control variables, we regard them as random variables with simple distribution functions.

Many interesting questions arise in connection with the planning of large-scale computations, particularly in relation to man–machine interactions. Some of these will be discussed in Volume III.

3.10. Stability Analysis

Let us now provide a simple stability analysis justifying the computational algorithm described above. The original functional equation is

$$f_k(c) = \min_y [h(c, y) + f_{k-1}(g(c, y))], \qquad k \geq 1,$$

$$f_0(c) = \min_y [h(c, y)]. \tag{3.10.1}$$

The calculation based upon discretization produces a sequence of functions $\{F_k(c)\}$ satisfying the relation

$$F_k(c) = \min_y [h(c, y) + F_{k-1}(g(c, y))] + e_k(c), \tag{3.10.2}$$

where $e_k(c)$ is an expression of the cumulative error due to the operations enumerated in Section 3.8. Our aim is to obtain a bound for $|f_k(c) - F_k(c)|$ in terms of an estimate for $|e_k(c)|$.

Let $y(c, k)$ be as before a function yielding the minimum in (3.10.1), an optimal policy, and $z(c, k)$ a corresponding function for (3.10.2). Then, by definition of these functions, we have

$$f_k(c) = h(c, y(c, k)) + f_{k-1}(g(c, y(c, k))) \leq h(c, z(c, k))$$
$$+ f_{k-1}(g(c, z(c, k))),$$
$$F_k(c) = h(c, z(c, k)) + F_{k-1}(g(c, z))) + e_k(c) \leq h(c, y(c, k)) \tag{3.10.3}$$
$$+ F_{k-1}(g(c, y(c, k))) + e_k(c).$$

It follows that

$$f_k(c) - F_k(c) \leq f_{k-1}(g(c, z)) - F_{k-1}(g(c, z)) - e_k(c)$$
$$\geq f_{k-1}(g(c, y)) - F_{k-1}(g(c, y)) - e_k(c). \tag{3.10.4}$$

(Here $z = z(c, k)$, $y = y(c, k)$.) Hence,

$$|f_k(c) - F_k(c)| \leq \max [|f_{k-1}(g(c, z)) - F_{k-1}(g(c, z))|,$$
$$|f_{k-1}(g(c, y)) - F_{k-1}(g(c, y))|] + |e_k(c)|. \tag{3.10.5}$$

Let

$$\varepsilon = \max_c |e_k(c)|. \tag{3.10.6}$$

Then from (3.10.5) we conclude that

$$\max_c |f_k(c) - F_k(c)| \leq \max_c |f_{k-1}(c) - F_{k-1}(c)| + \varepsilon, \tag{3.10.7}$$

whence

$$\max_c |f_k(c) - F_k(c)| \leq k\varepsilon. \tag{3.10.8}$$

This is a smooth enough growth of error to be called "stability." Naturally, we would prefer to have a bound which is independent of k. It is easy to see, however, that a bound which increases linearly with the number of stages means that the error cannot grow in a disastrous fashion. Acceptance of this kind of error estimate is common in the numerical analysis of ordinary and partial differential equations. The term "numerical instability" usually relates to an error term that grows exponentially with the number of stages.

EXERCISES

1. Consider the situation where $f_k(c) = \min_v [g(c, v) + f_{k-1}(h(c, v))]$, $\phi_k(c) = \min_w [g(c, w) + \phi_{k-1}(h_1(c, v))]$. Show that

   $$|f_k(c) - \phi_k(c)| \leq \max_v [|f_{k-1}(h(c, v)) - \phi_{k-1}(h_1(c, v))|]$$
   $$\leq \max_v [|f_{k-1}(h(c, v)) - \phi_{k-1}(h(c_1 v))|]$$
   $$+ \max_v [|\phi_{k-1}(h(c, v)) - \phi_{k-1}(h_1(c, v))|]$$

2. Let $\{f_k\}$, $k \geq 1$, be determined as above, $f_0(c) = \min_v g(c, v)$. Show that

   $$|f_k(c) - f_k(c')| \leq \max_v [|g(c, v) - g(c', v)| + |f_{k-1}(h(c, v)) - f_{k-1}(h(c', v))|]$$

3. Suppose that $\max_v |g(c, v) - g(c', v)| \leq a|c - c'|$, $\max_v |h(c, v) - h(c', v)| \leq b|c - c'|$. Let $|f_k(c) - f_k(c')| \leq u_k|c - c'|$, $k = 1, 2, \ldots, n$. Then $u_{n+1} \leq a + bu_n$.

4. Consider the case where $g(c, v) = g_1(c, v)\Delta$, $h(c, v) = c + h_1(c, v)\Delta$, where g_1 and h_p satisfy uniform Lipschitz conditions as above. Then u_n is bounded for $n = 1, 2, \ldots, N$, $N\Delta = T$ by a bound depending only on T and the Lipschitz constants.

5. Combining all of the foregoing consider the numerical stability of the scheme $f_k(c) = \min_v [g(c, v)\Delta + f_{k-1}(c + \Delta h(c, v))]$, $k \geq 1$, $f_0(c) = \min_v [g(c, v)\Delta]$.

3.11.   Functions and Solutions

The rapacious consumption of rapid-access storage involved in the foregoing treatment of multidimensional control processes using dynamic programming was due to the fact that we stored a function of R variables by means of its values at a rather dense set of points inside the R-dimensional unit cube. Specifically, we used $(M + 1)^R$ points, a number which increases painfully rapidly as both M and R increase.

Can we economize a bit? Clearly we cannot if we admit "arbitrary" functions, which is to say arbitrary correspondences between a grid point and a scalar value. However, we are not interested in arbitrary functions. The functions of primary concern to us possess "structure," which is to say there is a high degree of correlation between neighboring values.

Our task then is to utilize this structure in various ways which will enable us to reduce greatly the twin tasks of storage and retrieval. Many interesting and significant questions arise in this operation, questions of considerable algebraic and topological complexity. We will restrict our discussion here to a few basic aspects. In any case, not too much is known.

The aim is to be able to reconstruct the function $f_k(x)$, to some specified degree of accuracy, on the basis of considerably fewer than $(M + 1)^R$ values. One immediate approach is that of polynomial interpolation based on the assumption that the function of interest can be well approximated to by a polynomial. Observe that we are replacing the original purely numerical, or analytic, definition of the function by a mixed procedure, a set of numerical values together with an algorithm for producing functional values. In the previous sections we were perched at one extreme, that of a large set of functional values with no algorithms required, except possibly that of linear interpolation. In Chapter 5, on computational aspects of the calculus of variations in connection with two-point boundary-value problems, we will find it convenient to go to another extreme, that of virtually no numerical values, apart from a few parameters, but with complete dependence on algorithms.

What we are getting at is that we wish to employ a more general concept of "function" than that generally extant. We wish to conceive of a function not only in terms of the usual correspondence between two sets, the independent and dependent values, but far more importantly as a correspondence plus a particular algorithm for producing the set of values of the dependent variable given the set of values of the independent variable. We must specify not only the correspondence but also a procedure for realizing this correspondence.

Thus, we consider a function to be the *class* of all algorithms which can be used to produce the desired functional values. Which algorithm to use in a given situation depends critically upon a number of considerations of the type already discussed. The development of the computer has revived an operational point of view concerning the concept of function quite popular around the turn of the century; it was vigorously promoted earlier by Kronecker. For our present purposes, a function is of little interest unless we possess a feasible algorithm for obtaining numerical values.

Let us consider a simple example of this. The function e^t may be realized as the solution of the linear differential equation

$$u' - u = 0, \qquad u(0) = 1, \qquad (3.11.1)$$

as the limit of the sequence $\{u_N(t)\}$, where

$$u_N(t) = (1 + t/N)^N, \tag{3.11.2}$$

or as the limit of the sequence $\{v_N(t)\}$, where

$$v_N(t) = \sum_{k=0}^{N} t^k/k! . \tag{3.11.3}$$

It may also be determined from a table of values with a suitable interpolation formula, or by means of a suitable Padé approximation, and so on.

One of the consequences of the foregoing interpretation of the term "function" is that the terms "solution" and "problem" must also be interpreted operationally. Thus, e^t is a *solution* in most analytic contexts, but it is a *problem* if we wish to use a large set of its values in the course of a calculation. Rather than store 10^6 different values of e^t, we will include a simple algorithm for calculating it when needed.

Conversely, a differential equation such as

$$u'' + u + u^3 = 0 \qquad u(0) = 1, \qquad u'(0) = 0, \tag{3.11.4}$$

considered as a *problem* in most analytic contexts, is a *solution* when a digital computer is available and numerical values are desired over an interval equal in length to a period. The foregoing equation then represents a very simple set of instructions for producing functional values of $u(t)$. We shall use this fact strongly in Chapter 5.

The determination of a suitable algorithm is part of the control process involved in computing mentioned in Section 3.9.

EXERCISE

1. If we are required to solve $u'' - e^t u = 0$, $u(0) = 1$, $u'(0) = 0$, what are advantages and disadvantages of considering instead the system $u'' - vu = 0$, $u(0) = 1$, $u'(0) = 0$, $v' - v = 0$, $v(0) = 1$?

3.12.   Interpolation

Let us now turn to the storage and retrieval of functions. We begin with the case of a scalar function $u(t)$ specified at a set of points, $[t_1, t_2, \ldots, t_p]$, in $[0, 1]$. We wish to determine $u(t)$ at other points in the interval using polynomial interpolation. To do this, we can utilize the Lagrange interpolation formula,

$$u(t) = \sum_{i=1}^{p} \frac{u(t_i)v(t)}{(t - t_i)v'(t_i)}, \tag{3.12.1}$$

where

$$v(t) = \prod_{i=1}^{p} (t - t_i).$$ (3.12.2)

There are advantages and disadvantages (definite disadvantages if p is large) to a procedure of this type. We discuss some aspects of this in Section 3.14 and in Section 3.16.

EXERCISES

1. How do we determine the coefficients in $w(t) = \sum_{k=0}^{p} a_k t^k$ if we wish to minimize $\sum_{i=1}^{M} (w(t_i) - b_i)^2$, $M > p$?
2. How do we determine the coefficients a_0 and a_1 if we wish to minimize $\max_{0 \le t \le 1} |u(t) - a_0 - a_1 t|$?
3. What is the polynomial of degree $2p - 1$ which coincides with $u(t)$ at $t = t_i$, and whose derivative is equal to $u'(t)$ at $t = t_i$, where $i = 1, 2, \ldots, p$?

3.13. Computational Procedure

By cutting down drastically on the number of values stored, the interpolation technique reduces the total computing time in an equally drastic fashion. Let us briefly sketch the procedure. Let c be one-dimensional and suppose that we wish to determine $f_N(c)$ where

$$f_0(c) = \min_{y} h(c, y),$$

$$f_{k+1}(c) = \min_{y} [h(c, y) + f_k(g(c, y))], \qquad k \ge 0.$$ (3.13.1)

As above we suppose that $0 \le c \le 1$ and that we are employing some procedure to ensure that $0 \le g(c, y) \le 1$ for all admissible y.

We begin by calculating $f_0(c)$ at the p points $\{t_i\}$, using the equation

$$f_0(t_i) = \min_{y} h(t_i, y), \qquad i = 1, 2, \ldots, p.$$ (3.13.2)

The determination of $f_1(t_i)$ for $i = 1, 2, \ldots, p$ rests upon the equation

$$f_1(t_i) = \min_{y} [h(t_i, y) + f_0(g(t_i, y))].$$ (3.13.3)

To evaluate $f_0(g(t_i, y))$ for an admissible y-value, we employ the interpolation formula of (3.12.1), if needed, i.e., if $g(t_i, y) \ne t_j$. Having obtained in this way the values $\{f_1(t_i)\}$, we repeat the procedure to calculate the values $\{f_2(t_i)\}$, and so forth.

3.14. Evaluation of Polynomials

The use of interpolation formulas raises some questions interesting in their own right. The polynomial appearing in (3.12.1) will be evaluated a considerable number of times in the course of the calculation, namely ps times where p is the number of interpolation points and s the number of search points, the allowable values of y. Can we effect any saving of time in this operation?

Let us suppose that we have carried out the arithmetic operations required to write the interpolation polynomial in the form

$$L(t) = \sum_{k=0}^{p-1} a_k t^k. \tag{3.14.1}$$

The problem is then that of effective calculation of $L(t)$ for a particular value of t. To obtain a numerical value for $L(t)$, we can proceed as follows. We form the quantities $t^2, t^3, \ldots, t^{p-1}$, using the relation $t^k = t^{k-1}t$, a total of $p - 2$ multiplications, and then form the products $a_1 t, a_2 t^2, \ldots, a_{p-1} t^{p-1}$, another $p - 1$ multiplications. Thus we can evaluate $L(t)$ for a particular value of t using a total of

$$(p - 2) + (p - 1) = 2p - 3 \tag{3.14.2}$$

multiplications. Since multiplications consume about ten times as much time as additions, we shall ignore the p additions required. This is a straight-forward procedure, which means that with some thought we should do much better.

There is, indeed, a simple procedure, due to Horner, which requires only $p - 1$ multiplications. Write

$$L(t) = ((a_{p-1}t + a_{p-2})t + a_{p-3})t + \cdots . \tag{3.14.3}$$

It turns out that it is not possible to do better in general, but the proof is not a simple matter.

EXERCISES

1. Can one evaluate t^N in fewer than $N - 1$ multiplications?
2. Show that we can evaluate t^N in at most $2[\log_2 N]$ multiplications. (The minimum number of multiplications is not known for general N.)
3. What are some economical ways of evaluating a polynomial in two variables

$$p(t_1, t_2) = \sum_{k,l=1}^{M,N} a_{kl} t_1^{k} t_2^{l}?$$

4. What are some bounds on the number of multiplications required?

5. Can an arbitrary polynomial of degree $p - 1$ be written in the form $L(t) = L_1(L_2(t))$ where the degree of L_1 is m, the degree of L_2 is n, and $mn = p - 1$?

6. Can it be written in the form

$$L_1(L_2(t)) + L_3(L_4(t)) + \cdots + L_{2k-1}(L_{2k}(t))$$

for a suitable value of k? Is there a saving in evaluation time if $L(t)$ possesses this special form?

3.15.   Orthogonal Polynomials and Quadrature

We have considerable latitude with the set of interpolation points. Suppose that we make the selection on the basis of the following considerations. Let $u(t)$ possess an orthogonal expansion in terms of an orthonormal polynomial set over $[0, 1]$,

$$u(t) \sim \sum_{k=0}^{\infty} a_k \phi_k(t). \tag{3.15.1}$$

Since the Legendre polynomials, $\{p_n(t)\}$, are the orthogonal polynomials over $[-1, 1]$, the $\phi_k(t)$ are the suitably normalized versions of the shifted Legendre polynomials, $\{p_k(1 + t/2)\}$. The coefficients in (3.15.1) are determined by the relation

$$a_k = \int_0^1 u(t)\phi_k(t)\, dt. \tag{3.15.2}$$

To obtain a polynomial approximation to $u(t)$, we can employ the N-th partial sum,

$$u(t) \cong \sum_{k=0}^{N} a_k \phi_k(t). \tag{3.15.3}$$

To obtain a value for a_k from (3.15.2) we can employ a quadrature formula

$$a_k \cong \sum_{j=1}^{M} w_j u(t_j)\phi_k(t_j). \tag{3.15.4}$$

If we use a Gauss quadrature formula, we know that (3.15.4) is exact for $u(t)\phi_k(t)$, a polynomial of degree $2M - 1$ or less, and that the t_i are the M roots of $\phi_M(t) = 0$, all lying in $[0, 1]$. If then we take $M = N$, we have, for $k \neq l, 0 \leq k, l \leq N$,

$$0 = \int_0^1 \phi_k \phi_l\, dt = \sum_{j=1}^{N} w_j \phi_k(t_j)\phi_l(t_j). \tag{3.15.5}$$

This orthogonality relation suggests that we proceed directly. If the set $\{t_j\}$ is chosen to be the set of N assigned interpolation points (the roots of the shifted Legendre polynomial of degree N), we start from (3.15.5) and determine the coefficients a_k by means of the relation

$$a_k = \sum_{i=1}^{N} w_i u(t_i) \phi_k(t_i). \qquad (3.15.6)$$

Thus,

$$u(t) = \sum_{k=0}^{N} \phi_k(t) \left[\sum_{i=1}^{N} w_i u(t_i) \phi_k(t_i) \right]$$

$$= \sum_{i=1}^{N} w_i u(t_i) \left\{ \sum_{k=0}^{N} \phi_k(t) \phi_k(t_i) \right\}. \qquad (3.15.7)$$

EXERCISE

1. Using the familiar three-term recurrence relation satisfied by the $\phi_k(t)$, discuss ways of evaluating $u(t)$ as given in (3.15.7) which require fewer than the obvious number of multiplications.

3.16. Polygonal Approximation

In the foregoing sections, some of the difficulties involved in reconstructing a functional value have been discussed for the case where the function was represented by a polynomial of high degree. It is clear that some functions of quite simple form will require a polynomial of quite high degree to provide the required degree of accuracy if we insist upon using a polynomial interpolation formula. As an example of this, consider the case where the curve has the form shown in Figure 3.7. This is characteristic of a threshold and saturation effect.

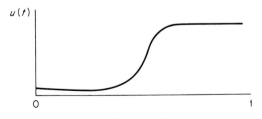

Figure 3.7

The flatter the beginning and end of the curve are, the higher the degree of the polynomial required to furnish a desired approximation. This in turn means a considerable increase in the time required to retrieve a value of $u(t)$.

Since this curve possesses a relatively simple shape, we should be able to devise a different type of approximation which permits a more rapid evaluation. For example, we may wish to employ a polygonal approximation, as in Figure 3.8. We write as the equations of the approximating function

$$u(t) = a_0 + b_0 t, \qquad 0 \le t \le t_1,$$
$$= a_1 + b_1 t, \qquad t_1 < t \le t_2,$$
$$= a_2 + b_2 t, \qquad t_2 < t \le 1. \tag{3.16.1}$$

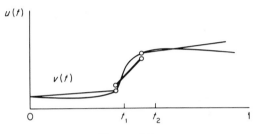

Figure   3.8

In each of the intervals $[t_i, t_{i+1}]$, $t_0 = 0$, $t_3 = 1$, the coefficients a_i and b_i are determined by the condition that they minimize some suitable measure of the degree of approximation. We can use a familiar criterion such as

$$\int_{t_i}^{t_{i+1}} (u(t) - a_i - b_i t)^2 \, dt, \tag{3.16.2}$$

or perhaps a Cebycev norm

$$\max_{t_i \le t \le t_{i+1}} |u(t) - a_i - b_i t| \tag{3.16.3}$$

if we are more ambitious.

Observe that in the foregoing figure we have carefully drawn the line segments so that they do not necessarily join at the transition points. We shall subsequently discuss the case where this linkage is required.

3.17.   Adaptive Polygonal Approximation

In the foregoing discussion it was tacitly assumed that the points t_1 and t_2 were fixed; see Figure 3.8. It is plausible that we can obtain an approximation which is significantly better if we allow these transition points to be initially variable, and subsequently determined by the curve itself. With this objective in mind, our aim is now to minimize the function

$$J(a_0, a_1, a_2, b_0, b_1, b_2, t_1, t_2) = \sum_{i=0}^{2} \int_{t_i}^{t_{i+1}} (u(t) - a_i - b_i t)^2 \, dt, \quad (3.17.1)$$

with respect to all of the parameters appearing. Here the restrictions on these parameters are

$$-\infty < a_0, a_1, a_2, b_0, b_1, b_2 < \infty, \qquad 0 \le t_1 \le t_2 \le 1. \quad (3.17.2)$$

As above, $t_0 = 0$, $t_3 = 1$.

If approached directly by means of calculus, this minimization leads to some unpleasant numerical problems, particularly if we allow a larger number of transition points.

EXERCISE

1. For the case where $u(t) = t^2$, determine the values of the a_i, b_i, and t_i first in the case where there is only one transition point and then, as above, when there are two.

3.18. Dynamic Programming Approach

Introduce the deviation function

$$\Delta(t_i, t_{i+1}) = \min_{a_i, b_i} \int_{t_i}^{t_{i+1}} (u(t) - a_i - b_i t)^2 \, dt, \quad (3.18.1)$$

defined for $0 \le t_i \le t_{i+1} \le 1$, and consider the more general question of minimizing the function

$$\sum_{i=0}^{N} \Delta(t_i, t_{i+1}), \quad (3.18.2)$$

over the region R defined by

$$0 = t_0 \le t_1 \le t_2 \le \cdots \le t_N \le t_{N+1} = a, \quad (3.18.3)$$

where $a \le 1$. We are supposing that $u(t)$ is defined only over $[0, 1]$.

Define the function

$$f_N(a) = \min_{R} \left\{ \sum_{i=0}^{N} \Delta(t_i, t_{i+1}) \right\}, \quad (3.18.4)$$

for $N = 0, 1, \ldots, 0 \le a \le 1$.

If we regard the choice of the t_i as a multistage process in which we choose first t_N, then t_{N-1}, and so on, we readily obtain the recurrence relation

$$f_N(a) = \min_{0 \le t_N \le a} [\Delta(t_N, a) + f_{N-1}(t_N)], \qquad N \ge 1, \qquad (3.18.5)$$

with the initial condition

$$f_0(a) = \Delta(0, a). \qquad (3.18.6)$$

This functional equation can be made the basis of a simple algorithm for determining the transition points, $t_1, t_2, \ldots, t_N$.

EXERCISES

1. Explain how the foregoing could be used as a pattern recognition scheme to distinguish between a V and a W.

2. Obtain the value of $\min_{a_i} \int_0^1 (t^N - a_0 - a_1 t - \cdots - a_{N-1} t^{N-1})^2 \, dt$ using Legendre polynomials.

3. Show that

$$\min_{a_i} \int_s^r (t^M - a_0 - a_1 t - \cdots - a_{N-1} t^{M-1})^2 \, dt$$

$$= \min_{a_i} \int_0^{r-s} (t^M - a_0 - a_1 t - \cdots - a_{N-1} t^{M-1})^2 \, dt, \quad r \ge s,$$

and that

$$\min_{a_i} \int_0^b (t^M - a_0 - a_1 t - \cdots - a_{N-1} t^{M-1})^2 \, dt$$

$$= b^{M+1} \min_{a_i} \int_0^1 (t^M - a_0 - a_1 t - \cdots - a_{N-1} t^{M-1})^2 \, dt.$$

4. Using the foregoing results, evaluate $f_N(a)$ in the case where $u(t) = t^M$, $M \ge 0$.

5. What are corresponding results for the case where $u(t) = e^{ct}$?

6. Evaluate $f_N(a)$ in the case where $u(t) = t^2$ and we employ the Cebycev norm

$$\Delta(t_i, t_{i+1}) = \min_{a_i} \max_{t_i \le t \le t_{i+1}} |t^2 - a_0 - a_1|.$$

7. Obtain a corresponding algorithm for the case where it is desired to minimize $\max_{0 \le i \le N} \Delta(t_i, t_{i+1})$.

8. Can we treat the problem of determining optimal variable step size in the numerical integration of ordinary differential equations using the foregoing methods?

9. How do we treat the problem where the function $u(t)$ is replaced by a set of functional values $\{u(k\Delta)\}$?

3.19. Linkage

If we impose the further condition that the set of approximating line segments must itself be a continuous function, which is to say that the individual segments must be joined at the transition points, t_i, a slight modification of the foregoing is necessary. We begin by introducing a function of two variables

$$f_N(a, b) = \min_{t_i} \sum_{i=0}^{N} \Delta(t_i, t_{i+1}), \tag{3.19.1}$$

where the t_i are subject to the constraint of continuity just mentioned, and we impose the condition that the approximating curve assumes the value b at a. Let us speak about straight line approximation for the sake of simplicity,

$$v(t) = a_N + b_N t, \qquad t_N \leq t \leq a. \tag{3.19.2}$$

Then

$$\Delta(t_N, a) = \min_{a_N, b_N} \int_{t_N}^{a} (u(t) - a_N - b_N t)^2 \, dt, \tag{3.19.3}$$

where the minimum is over values of a_N and b_N satisfying $a_N + b_N a = b$.
The basic recurrence relation is

$$f_N(a, b) = \min_{t_N} [\Delta(t_N, a) + f_{N-1}(t_N, a_N + b_N t_N)], \tag{3.19.4}$$

where a_N and b_N are determined as functions of t_N and a by the minimization in (3.19.3). Once $f_N(a, b)$ has been determined by means of (3.19.4), we complete the solution by minimizing this function with respect to b.

EXERCISE

1. Determine $f_N(a, b)$ for the case where $u(t) = t^2$.

3.20. Search Processes

In the previous sections we focussed attention on some aspects of the task involved in utilizing the structure of the individual function to reduce the effort involved in the storage and retrieval of data. It is to be expected that the process of searching for a minimum can also be facilitated by the use of various types of information concerning the structure of the function under consideration.

A structure of paramount importance is that of unimodality, which is to say the function possesses only one relative minimum, necessarily the absolute minimum; see Figure 3.9.

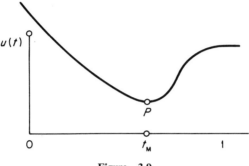

Figure   3.9

Suppose that we wish to locate the point t_M at which the function assumes its minimum value to an accuracy of say one in 10^6. We can do this, of course, in a rather brutal direct fashion by dividing the interval $[0, 1]$ into intervals of length 10^{-6} and laboriously comparing the functional values, point after point. This requires, on the average, $10^6/2$ evaluations, assuming that there is equal probability of the point t_M lying anywhere in $[0, 1]$.

Let us proceed instead in the following, craftier fashion. Divide the interval $[0, 1]$ into four sub-intervals $[0, 1/4]$, $[1/4, 1/2]$, $[1/2, 3/4]$, $[3/4, 1]$ and evaluate the function at the three points $1/4$, $1/2$, $3/4$. Examining the various possibilities, we see that by means of these three evaluations and two comparisons, we can restrict the position of the minimum value to within an interval of length one-half the original interval. With the aid of two more evaluations and two more comparisons, the point t_M can be restricted to an interval of length one-quarter of the original interval. Thus, with the examination of $2k + 1$ points, we can narrow the interval containing t_M to one of length 2^{-k}. If we wish

$$2^{-k} < 10^{-6}, \tag{3.20.1}$$

we must take

$$k \geq \log (10^6)/\log 2. \tag{3.20.2}$$

This number represents a vast improvement over $10^6/2$.

How well can we do with an optimal search technique, and what do we mean by "optimal search?" This area, the theory of search processes, is a new and difficult one with many important and entertaining, unresolved problems. A number of references will be found at the end of this chapter.

3.21. Discussion

What is particularly interesting about the foregoing is that an analysis of the effort required in a computational solution of some typical functional equations of dynamic programming discloses the existence of many further classes of dynamic programming processes. A simple way of generating these processes is to postulate that there is nothing routine about any type of numerical analysis even when it is known that the algorithm is effective according to various criteria. There always remain the problems of reducing the time required to obtain a specified accuracy, of maximizing the accuracy obtainable in a given time, and of using various combinations of algorithms in connection with various computers.

The situation is much the same when we turn to the writing of the actual computer program. There are many intricate questions of scheduling and allocation of effort connected with the internal operation of the hybrid computer, some of which themselves can be treated by means of dynamic programming methods.

3.22. Continuity

If we impose further restrictions on the nature of the minimand, which is to say on its analytic and geometric structure, we can do even better. Let us employ the powerful tool of continuity. Consider the familiar equation

$$f_{k+1}(c) = \min_{v} [h(c, v) + f_k(g(c, v))], \tag{3.22.1}$$

where we replace y by v to indicate the scalar nature of the problem chosen for illustrative purposes. Suppose that we can establish the fact that the minimand is convex in v, as in Figure 3.10. Then, as we change the value of c from c_1 to a neighboring value c_2, the position of the minimizing point will

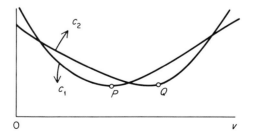

Figure 3.10

change slowly. This means that it is sufficient to search in the neighborhood of P, the minimizing point for c_1, in order to locate Q.

This is a useful idea even if the minimand is not convex. Consider, for example, a situation where there are only two local minima, as in Figure 3.11.

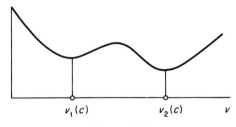

$v_1(c)$ $v_2(c)$ v

Figure   3.11

What can happen is that as c varies, the absolute minimum can shift from the neighborhood of $v_1(c)$ to the neighborhood of $v_2(c)$. Nonetheless, by keeping track of both the absolute minimum and the second minimum, we can use the property of continuity to reduce the search time considerably. A related idea is to use the minimizing value $v(c, N - 1)$ computed and stored for the previous stage as a first guess for the minimizing value $v(c, N)$ for the current stage.

EXERCISE

1.   Discuss the feasibility of this idea.

3.23.   Restriction to Grid Points

Previously we have described the use of interpolation techniques of various degrees of sophistication to evaluate $f_k(g(c, y))$ when $g(c, y)$ does not coincide with one of the assigned grid points. One way to avoid the problems arising from interpolation is to introduce a constraint on y, thereby forcing it to assume only values which will put $g(c, y)$ at a grid point.

Thus, for example, if we are trying to determine an optimal trajectory from P to Q, we can begin with a grid of the form in Figure 3.12.

From P we must go to one of the three nearby points; from whichever of these points we select, we can only go to one of three nearby points to the right, and so on.

In Chapter 5, we shall describe in detail how we can use this approximate trajectory to obtain a far better approximation, and why we are forced to resort to this technique in many cases. In Chapter 10 we discuss generalized routing problems of this nature.

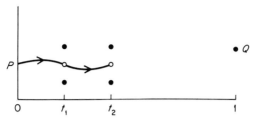

Figure   3.12

3.24.   Multidimensional Case

We have refrained from any discussion of the associated multidimensional problems involved in storage and retrieval and in search because so little has been done in this area, due to the intrinsic difficulty of the problems. Some references will be found at the end of this chapter.

Miscellaneous Exercises

1. Consider the problem of maximizing $a_1 x_N + a_2 y_N$ where $x_{n+1} = a_{11} u_n + a_{12} v_n$, $x_0 = c_1$, $y_{n+1} = a_{21} u_n + a_{22} v_n$, $y_0 = c_2$, and $0 \le u_n \le x_n$, $0 \le v_n \le y_n$. If $f_N(c_1, c_2) = \max [a_1 x_N + a_2 y_N]$, show that $f_N(c_1, c_2) = \max_R f_{N-1}(a_{11} u + a_{12} v, a_{21} u + a_{22} v)$, where R is defined by $0 \le u \le c_1$, $0 \le v \le c_2$, and that $f_N(c_1, c_2) = c_1 f_N(1, c_2/c_1) = c_2 f_N(c_1/c_2, 1)$.

2. Hence, show that by introducing the *two* functions $f_N(1, t) = g(t)$, $f_N(t, 1) = h(t)$, $0 \le t \le 1$, we can calculate $f_N(c_1, c_2)$ for general values of $c_1, c_2 \ge 0$, in terms of functions of one variable without worrying about an increasing grid.

    (Problems of this nature arise in the study of "bottleneck processes;" see the books by Bellman and Bellman-Dreyfus on dynamic programming.)

3. Show that we can employ the same device to handle the problem of minimizing $\sum_{n=0}^{N} [(x_n, x_n) + (y_n, y_n)]$ subject to $x_{n+1} = Ax_n + y_n$, $x_0 = c$, $x_n \ge 0$, where this last condition means that all components are nonnegative. (This is an important constraint in many applications of economic and biomedical origin.)

4. The problem of determining the smallest characteristic value of $u'' + \lambda \phi(t) u = 0$, $u(0) = u(1) = 0$, $\phi(t) \ge a_1 > 0$, can be approximated by the problem of minimizing $J_N = \sum_{k=1}^{N} (u_k - u_{k-1})^2 \Delta$ subject to $\sum_{k=1}^{N} \phi_k u_k^2 \Delta = 1$, $u_0 = u_N = 0$. Imbed this within the family of problems where $u_0 = b$. Obtain a functional equation for $\min J_N$.

5. Discuss the computational feasibility of this approach, and, particularly, the sensitive nature of the value $b = 0$.

6. Can one obtain similar functional equations for the higher characteristic values?

BIBLIOGRAPHY AND COMMENTS

3.1. For a number of applications and further discussion, let us refer to the books

R. Bellman and S. Dreyfus, *Applied Dynamic Programming*, Princeton Univ. Press, Princeton, New Jersey, 1962.

R. Larson, *State Increment Dynamic Programming*, Elsevier New York, 1968.

A number of extensions of the older algorithms, as well as new algorithms, are being currently developed. See the book by Larson cited above and

U. Bertele and F. Brioschi, "A New Algorithm for the Solution of the Secondary Optimization Problem in Nonserial Dynamic Programming," *J. Math. Anal. Appl.* **27**, 1969,

For additional applications, see

R. Aris, G. L. Nemhauser, and D. J. Wilde, "Optimization of Multistage Cyclic and Branching Systems by Serial Procedures," *A.I.Ch.E. J.* **10**, 1964, p. 913.

P. A. Gilmore, "Structuring of Parallel Algorithms," *J. Assoc. Comp. Mach.*, **15**, 1968, pp. 176–192.

R. E. Larson and W. G. Keckler, "Comments on Interpolation and Extrapolation Schemes in Dynamic Programming," *IEEE Trans. Automat. Control*, AC-3, 1968, pp. 294–296.

N. N. Moiseev, "Methods of Dynamic Programming in the Theory of Optimal Controls (IV), "*USSR Computa. Math. Math. Phys.*, **5**, 1965, pp. 58–75.

C. J. Rose, "Dynamic Programming Processes within Dynamic Programming Processes," *JMAA*,

D. J. Wilde, and C. S. Beightler, *Foundations of Optimization*, Prentice Hall, New York, 1967.

P. J. Wong, *Reducing the Computational Time of Dynamic Programming*, to appear.

P. J. Wong and D. G. Luenberger, "Reducing the Memory Requirements of Dynamic Programming," *Oper. Res.* **16**, 1968, pp. 1115–1125.

P. J. Wong, "A New Decomposition Procedure for Dynamic Programming," *Oper. Res.*, **18**, 1970, pp. 119–131.

3.4. There are many interesting and subtle questions connected with the storage and retrieval of functions, See

G. G. Lorentz, "Metric Entropy and Approximation," *Bull. Amer. Math. Soc.*, **72**, 1966, pp. 903–937,

where the work of Kolmogoroff and his associates is discussed. Many references will be found there. See also

A. R. Butz, "Convergence with Hilbert's Space Filling Curve," *J. Computer Sys. Sci.*, **3**, 1969, pp. 128–146.

For some use of the structure to obtain a computationally feasible algorithms, see

R. Bellman, "On the Computational Solution of Linear Programming Problems Involving Almost-Block-Diagonal Matrices," *Manag. Sci.*, 3, 1957, pp. 403–406.

S. Beightler and D. B. Johnson, "Superposition in Branching Allocation Problems," *JMAA*, 12, 1965, pp. 65–70.

G. L. Nemhauser and Z. Ullmann, "Discrete Dynamic Programming and Capital Allocation," *Manag. Sci.* 15, 1969, pp. 494–505.

P. J. M. van den Bogaard, A. M. Luque and C. Van de Panne, "A Study of the Implications of Alternative Layers in Quadratic Dynamic Programming," *Rev. Fr. de Recherche Operationelle*, 6, 1962, pp. 163–183.

We have not discussed any of the interesting problems connected with the use of successive approximations to reduce the dimension of the functions, that occur. See, however, some of the later chapters where work of Angel, Collins and Lew is cited.

3.5.   In situations in which it is desired to calculate the solution of $f = T(f)$ by means of an iterative scheme of the form $f_{n+1} = T(f_n)$, convergence can often be greatly accelerated by means of a nonlinear summability method. See

R. Bellman and R. Kalaba, "A Note on Nonlinear Summability Techniques in Invariant Imbedding," *J. Math. Appl.* 6, 1963, pp. 465–472, for an application of this idea and further references.

For a description of a new computer on the horizon, see

D. J. Kuck, "Illiac IV Software and Application Programming," *IEEE Trans. Computers*, C-17, 1968, pp. 758–770.
3.7.   See the books referred to in 3.1 for further discussions of this basic idea.
3.8.   See

R. Bellman, *Methods of Nonlinear Analysis*, Academic Press, New York, 1970,

for a detailed discussion of the general concept of stability, and

J. J. G. Guignabodet, "Majoration des Erreurs de Quantification dans les Calculs de Programmation Dynamique," *Compt. Rend.*, 255, 1962, pp. 828–830.

A. Korsak, *Perturbed Optimal Control Problems*, Dept. of Mathematics, Univ. of California, Berkeley, 1966.
3.9.   See

R. E. Kalman, "Towards a Theory of Difficulty of Computation in Optimal Control," *Proc. IBM Sci. Symp. Control Theory Appl.* 1964.

J. Todd, "Motivation for Working in Numerical Analysis," *Comm. Pure Appl. Math.*, 8, 1955, pp. 97–116.

J. Todd, "The Problem of Error in Digital Computation," *Error Digital Computat*, 1, 1965, pp. 3–41.

I. Babushka and S. L. Sobolev, "Optimization of Numerical Processes" (in Russian), *Appl. Math.* 10, 1965, pp. 96–129. (See *Math. Rev.*, no. 8481, 32, 1966).
3.11.   See

R. Bellman, "A Function is a Mapping—Plus a Class of Algorithms," *Inform. Sci.* to appear.
3.13.   An interesting question is whether or not it is better to use a set of low degree approximations over small regions or one high degree approximation over the entire region. See

S. P. Azen, *Successive Approximations by Quadratic Fitting as Applied to Optimization Problems*, The RAND Corp., RM-5001-PR, 1966.
3.14. See

V. Ja Pan, "On Means of Calculating Values of Polynomials" (in Russian), **21**, 1966, pp. 103–134. (See *Math. Rev.*, no. 6994, **34**, 1967.

A. Ostrowski, "On Two Problems in Abstract Algebra Connected with Horner's Rule," " *Studies in Mathematics and Mechanics Presented to Richard von Mises*, Academic Press, New York, 1954, pp. 40–48.

A. Schronhage, "Multiplication grosser Zahlen," Computing (*Arch. Elektron, Rechnen*), **1**, 1966, pp. 182–196.

F. J. Smith, "An Algorithm for Summing Orthogonal Polynomial Series and Their Derivatives with Applications to Curve-Fitting and Interpolation," *Math. Comp.*, **19**, 1965, pp. 33–36.
3.15. See

R. Bellman and R. Kalaba, *Quasilinearization and Nonlinear Boundary-value Problems*, Elsevier, New York, 1965.

H. B. Keller, "On the Pointwise Convergence of the Discrete-ordinate Method," *J. Soc. Indus. Appl. Math.*, **8**, 1960, pp. 560–567.
3.16. See

M. Aoki, "On the Approximation of Trajectories and its Application to Control Theory Optimization Problems," *JMAA*, **9**, 1964., pp. 23–41.

R. Bellman, "On the Approximation of Curves by Line Segments Using Dynamic Programming," *Comm. Assoc. Comput. Machinery*, **4**, 1961, pp. 284.

R. Bellman, "Segmental Differential Approximation and the 'Black Box' Problem," *J. Math. Anal. Appl.*, **12**, 1965, pp. 91–104.

R. Bellman and R. Roth, "Segmental Differential Approximation and Biological Systems: An Analysis of a Metabolic Process," *J. Theoret. Biol.*, **11**, 1966, pp. 168–176.

R. Bellman, and R. Roth, "A Technique for the Analysis of a Broad Class of Biological Systems," *Bionics Symp.*, pp. 725–737.

R. Bellman, B. Gluss, and R. Roth, "On the Identification of Systems and the Unscrambling of Data: Some Problems Suggested by Neurophysiology," *Proc. Nat. Acad. Sci. USA*, **52**, 1964, pp. 1239–1249.

B. Gluss, "Least Squares Fitting of Planes to Surfaces Using Dynamic Programming," *Comm. Assoc. Computing Machinery*, **6**, 1963, pp. 172–175.

B. Gluss, "An Alternative Method for Continuous Line Segment Curve-Fitting," *Inform. Control*, **7**, 1964, pp. 200–206.

A. G. Lubowe, "Optimal Functional Approximation Using Dynamic Programming," *AIAA J.*, **2**, 1964, pp. 376–377.

R. Roth, "The Unscrambling of Data, Studies in Segmental Differential Approximation," *J. Math. Anal. Appl.*, **14**, 1966, pp. 5–22.

H. Stone, "Approximation of Curves by Line Segments," *Math. of Comp.*, **15**, 1961, pp. 40–47.

A. Wouk, "Approximation and Allocation," *J. Math. Anal. Appl.*, **8**, 1964, pp. 135–143.

In some of the papers approximation problems are considered in which the straight line segment, a solution of $u'' = 0$, is replaced by the solution of a general class of linear or nonlinear differential equations.

3.20. For a number of further results and references concerning search processes, see the book by Bellman and Dreyfus referred to in 3.1. See also

M. Avriel and D. J. Wilde, "Golden Block Search for the Maximum of Unimodal Functions, *Manag. Sci.*, **14**, 1968, pp. 307–319.

W. W. Chu, *Optimal Adaptive Search*, Tech. Rp. 6252-1, Stanford Electronics Laboratories, 1966.

P. Krolak and L. Cooper, "An Extension of Fibonaccian Search to Several Variables," *Comm. ACM.*, **6**, 1963, pp. 369–641.

J. Sugie, "An Extension of Fibonaccian Searching to the Multidimensional Cases," *IEEE Trans. Automat. Control*, **AC-9**, Jan. 1964.

    3.24.   See the references in 3.1, particularly the book by Wilde.

Some papers discussing alternate definitions of state are

R. Bellman, "On the Reduction of Dimensionality for Classes of Dynamic Programming Processes," *J. Math. Anal. Appl.*, **23**, 1961, pp. 358–360.

R. Bellman, "Dynamic Programming, Generalized States and Switching Systems," *J. Math. Anal. Appl.*, **12**, 1965, pp. 360–363.

4

CONTINUOUS CONTROL PROCESSES
AND THE CALCULUS OF VARIATIONS

4.1. Introduction

In this and the following chapter we wish to consider some fundamental aspects of the application of the calculus of variations to the study of continuous control processes. In this chapter we shall concentrate on the analytic details, reserving Chapter 5 for computational questions.

Many control processes lead to the problem of minimizing the functional

$$J(x, y) = \int_0^T h(x, y) \, dt \qquad (4.1.1)$$

with respect to the vector function $y(t)$ where the vectors x and y are related by means of the differential equation

$$x' = g(x, y), \qquad x(0) = c. \qquad (4.1.2)$$

In some cases, there are terminal conditions of the form

$$m(x(T), x'(T)) = 0, \qquad (4.1.3)$$

and in others a terminal term appears in the criterion function,

$$J(x, y) = k(x(T)) + \int_0^T h(x, y) \, dt. \qquad (4.1.4)$$

In addition, there may be a global condition of the type

$$\int_0^T r(x, y) \, dt \le a_1. \tag{4.1.5}$$

In economic and engineering processes this often represents a limitation on resources.

We will steadfastly avoid any discussion of local constraints on either the state or control variables, e.g., conditions of the form $\|x\| \le a_2$, $\|y\| \le a_3$. Although control processes involving constraints are of major importance in applications, they require a bit more sophisticated analysis. Hence, we will defer their consideration to a separate volume.

To illustrate a number of the ideas involved in as simple a setting as possible, we will concentrate initially on the minimization of functionals of the particular form

$$J(u) = \int_0^T (u'^2 + 2h(u)) \, dt, \tag{4.1.6}$$

where u is a scalar function subject to $u(0) = c$. At the close of the chapter we will consider some multidimensional versions.

As in Volume I, two basic approaches will be employed: One constructive and one existential, based on functional analysis. The first method hinges on the use of the Euler equation, the principal tool of the calculus of variations. It permits us to assert existence and uniqueness of the minimizing function in a local neighborhood of function space. The second method enables us to treat global regions. Each method requires its own set of auxiliary restrictive conditions. This lack of uniformity and generality must be endured with some degree of equanimity because we neither possess, nor anticipate, a global theory of nonlinear processes.

Since we are interested only in exhibiting some of the ways in which the calculus of variations can be applied to important classes of continuous control processes, we make no incursion into this classical field that is not specifically needed for our immediate purposes. References to excellent contemporary texts in the realm of the calculus of variations (some specifically aimed at control theory) will be found at the end of this chapter.

Our major mechanism for handling the Euler equation will be the method of successive approximations, the usual general factotum. In this chapter this basic tool will be applied in a rather humdrum and direct fashion. In Chapter 5, where computational solution is our goal we will be forced to use more sophistication. We have throughout emphasized this point that effective determination of numerical results requires more analytic know-how than demonstration of existence and uniqueness. Our modest objectives in this chapter are to lay a rigorous foundation for the methods of the succeeding chapter and to provide some needed results for the chapter succeeding that in which dynamic programming is applied to continuous control processes.

4.2.   The Quadratic Case

The path we follow in the nonlinear case will perhaps appear more plausible if we first briefly review our previous steps in the far simpler case where the criterion functional is quadratic. These matters were discussed at length in Volume I.

To be specific, let us consider the minimization of the quadratic functional

$$J(u) = \int_0^T (u'^2 + u^2)\, dt \tag{4.2.1}$$

over the class of functions satisfying the initial condition $u(0) = c$, and the further condition that $\int_0^T u'^2\, dt < \infty$. We write $u' \in L^2(0, T)$ to indicate this latter condition. A function satisfying both of these conditions will be called *admissible*. An admissible function of particular form is often called a *trial function*.

Assume for the moment that the minimum exists and that u is a minimizing function. Consider $J(u + \varepsilon v)$ where v is such that v' also belongs to $L^2(0, T)$, ε is a scalar parameter, and $v(0) = 0$. Hence, $u + \varepsilon v$ is an admissible function. We have

$$J(u + \varepsilon v) = \int_0^T [(u' + \varepsilon v')^2 + (u + \varepsilon v)^2]\, dt$$

$$= J(u) + \varepsilon^2 J(v) + 2\varepsilon \int_0^T [u'v' + uv]\, dt. \tag{4.2.2}$$

Since, by assumption $J(u)$ is the minimum value, we must have the variational condition

$$\int_0^T [u'v' + uv]\, dt = 0. \tag{4.2.3}$$

Let us blithely assume that u' is differentiable and integrate the term $u'v'$ by parts. Nothing in the original set of assumptions guarantees the legitimacy of this operation. We have supposed only that $u' \in L^2(0, T)$. Nonetheless, it costs very little to proceed formally and see what happens. The result is

$$\left[u'v\right]_0^T + \int_0^T v[-u'' + u]\, dt = 0. \tag{4.2.4}$$

If this equation is to hold for all v, such that $v' \in L^2(0, T)$, it is *plausible* that a consequence is the equation

$$-u'' + u = 0, \tag{4.2.5}$$

and an additional condition

$$u'(T) = 0, \tag{4.2.6}$$

a boundary condition.

The differential equation is the celebrated *Euler equation* associated with the foregoing variational problem.

At this point we abruptly reverse our gears and start with an investigation of the equation in (4.2.5) subject to (4.2.6) and $u(0) = c$. A simple and direct calculation shows that there is a unique solution to this equation. Furthermore, it is easily seen that this function furnishes the absolute minimum of J.

For any other admissible function w, we have

$$J(w) = J(u + (w - u)) = J(u) + J(w - u)$$
$$+ 2 \int_0^T [(w' - u')u' + (w - u)u] \, dt. \tag{4.2.7}$$

Integrating the third term by parts, we see that it vanishes by virtue of the Euler equation satisfied by u. Hence,

$$J(w) = J(u) + J(w - u) > J(u) \tag{4.2.8}$$

if $w \neq u$.

If we consider a more general problem where the criterion function is

$$J(u) = \int_0^T (u'^2 + q(t)u^2) \, dt, \tag{4.2.9}$$

we derive the corresponding Euler equation

$$u'' - q(t)u = 0, \qquad u(0) = c, \qquad u'(T) = 0, \tag{4.2.10}$$

in a similar fashion. Since we cannot explicitly solve equations of this nature in general, we must use simpler and more general tools to obtain our desired results. These extend readily to higher dimension. The basic hypothesis is the positive definiteness of $J(u)$ for a suitable class of u. See Volume I for the details.

4.3. Discussion

In the treatment of the minimization of the quadratic functional

$$J(u) = \int_0^T (u'^2 + q(t)u^2) \, dt, \tag{4.3.1}$$

the linearity of the Euler equation plays a crucial role. For general functionals with nonlinear Euler equations, we shall be forced to work harder to achieve less. Nonetheless, with a bit of perseverance and sufficient restriction of generality, we can carry out a general plan consisting of the following steps:

(a) formal derivation of the Euler equation;
(b) demonstration of the existence of a solution of the Euler equation;
(c) demonstration of conditional uniqueness of this solution;
(d) demonstration of conditional minimization of the original functional.

In some cases the control theory background enables us to go a bit further since we can invoke the nonnegativity of certain "costs."

4.4. Formal Derivation of the Euler Equation

Let us, as promised, use the problem of minimizing the relatively simple functional

$$J(u) = \int_0^T (u'^2 + 2h(u))\, dt \tag{4.4.1}$$

to illustrate some general methods. We suppose that T is finite and that

(a) $u(0) = c$,

(b) $\displaystyle\int_0^T u'^2 \, dt < \infty,$ $\tag{4.4.2}$

(c) $h'(u)$ and $h''(u)$ exist for all u.

In many cases, h is a polynomial, automatically ensuring (2c). Let us follow the general procedure outlined in Section 4.3. Let u be a, or the, minimizing function and consider the expression

$$J(u + \varepsilon v) = \int_0^T [(u' + \varepsilon v')^2 + 2h(u + \varepsilon v)] \, dt, \tag{4.4.3}$$

where v is another function satisfying the conditions of (4.4.2b) with $v(0) = 0$; ε is a scalar.

If $f(\varepsilon) = J(u + \varepsilon v)$ is to possess a minimum at $\varepsilon = 0$, then the following condition holds,

$$f'(0) = \frac{d}{d\varepsilon} J(u + \varepsilon v)\Big|_{\epsilon = 0} = 0. \tag{4.4.4}$$

Hence, we must have

$$\int_0^T [u'v' + h'(u)v] \, dt = 0, \tag{4.4.5}$$

for all v such that $v' \in L^2(0, T)$, $v(0) = 0$.

Integrating by parts, under the assumption, as yet unsubstantiated, that the minimizing function u possesses a second derivative, we obtain

$$u'(T)v(T) + \int_0^T v[-u'' + h'(u)]\, dt = 0. \tag{4.4.6}$$

Thus, we are led to the Euler equation

$$u'' - h'(u) = 0,$$
$$u(0) = c, \qquad u'(T) = 0. \tag{4.4.7}$$

The original variational problem has thus been transformed into the problem of solving a nonlinear differential equation subject to a two-point boundary-value condition—or has it?

4.5. Haar's Device

As mentioned above, the Euler equation possesses a flimsy claim to validity at the moment because we have no guarantee that u'' exists. Let us show, nevertheless, following Haar, that it is quite easy to derive (4.4.7) from (4.4.5) in a rigorous fashion. From the standpoint of classical analysis the procedure we present unfortunately looks like an ad hoc device; from the standpoint of dynamic programming, where u' is a more natural function than u, it is a simple and logical step.

Turning to (4.4.5), let us regard v' as the basic variable and integrate the term $h'(u)v$ by parts to eliminate v. The result is

$$\int_0^T h'(u)v\, dt = \left[-\left(\int_t^T h'(u)\, dt_1\right)v\right]_0^T + \int_0^T \left(\int_t^T h'(u)\, dt_1\right)v'\, dt. \tag{4.5.1}$$

We integrate from t to T so that both terms in the integrated part drop out; recall that $v(0) = 0$. Hence, (4.4.5) leads to

$$\int_0^T v'\left[u' + \int_t^T h'(u)\, dt_1\right] dt = 0, \tag{4.5.2}$$

for all $v' \in L^2(0, T)$. If we now choose

$$v' = u' + \int_t^T h'(u)\, dt_1, \qquad v(0) = 0, \tag{4.5.3}$$

the expression in (4.5.2) leads to

$$\int_0^T \left[u' + \int_t^T h'(u)\, dt_1\right]^2 dt = 0, \tag{4.5.4}$$

whence

$$u' + \int_t^T h'(u)\, dt_1 = 0 \tag{4.5.5}$$

almost everywhere. The procedure is legitimate if we can show that $u' +$
$\int_t^T h'(u)\, dt_1$ belongs to $L^2(0, T)$ and thus qualifies as an admissible v'. But this
is clear since $u' \in L^2(0, T)$ by assumption, and $\int_t^T h'(u)\, dt_1$ is a continuous
function in $[0, T]$.

Since (4.5.5) holds almost everywhere, we see that u' is equal to a continuous
function almost everywhere. We might just as well extend u' by continuity
and have (4.5.5) hold for all t in $[0, T]$. From this it follows that u' obtained
in this way possesses a derivative and thus that u satisfies the Euler equation
of (4.4.7).

EXERCISES

1. Show that the Euler equation associated with $\int_0^T g(u', u)\, dt$ is $(g_{u'})' = g_u$.
2. Under what assumptions concerning g can we use Haar's device to obtain
   the Euler equation?
3. Obtain the Euler equation formally by considering the limit of the
   variational conditions holding for the minimization of $\sum_{n=0}^N (u_{n+1} - u_n)^2$
   $+ 2 \sum_{n=0}^N h(u_n)$, $u_0 = c$ (Euler).

4.6.   Discussion

The foregoing derivation is interesting and important, but not particularly
significant at the moment since we are pledged to follow the route outlined
in Section 4.3. Our plan is to regard the Euler equation as a Deus ex Machina
and to show first that for sufficiently small T the equation possesses a unique
solution within a class of acceptable functions. Secondly, we shall demonstrate that the function so obtained furnishes an absolute minimum of $J(u)$
provided that we suitably restrict the class of admissible functions.

We shall furnish constructive techniques for bounding T and determining u.
All of this, it must be admitted, is under the condition that T is small, which
means that we are forging a *local* rather than *global* theory. The fact that the
calculus of variations is basically a local theory in function space is one of its
principal drawbacks. We shall discuss this point again. It corresponds to the
difficulty that one faces in applying calculus to the minimization of a function
of a finite number of variables.

In fairness, however, to this venerable theory, it must be added, and emphasized, that no uniformly powerful methods exist for treating control processes.
Each of the existing theories and techniques performs well in connection with
certain carefully chosen problems. A problem chosen at random usually
embarasses everybody. So much the worse for parochialism—so much the
better for mathematics!

4.7. Nonexistence of a Minimum—I

Prior to a discussion of the Euler equation of (4.4.7) along the lines just indicated, let us present three simple examples to illustrate that we are not being unreasonably fussy in spending some time establishing existence and uniqueness of minimizing functions in appropriate function classes. To begin with, consider the simple functional

$$J(u) = \int_0^T u^2 \, dt, \tag{4.7.1}$$

with $u(0) = c \neq 0$. If we seek the minimum of $J(u)$ in the class of functions for which J exists, namely $u \in L^2(0, T)$, it is clear that the minimum value is zero attained for the function $u = 0$ almost everywhere.

On the other hand, if we seek the minimum in the narrower class of continuous functions, it is clear that a minimum value of zero is *not* attained. There is no continuous function u over $[0, T]$ satisfying $u(0) = c \neq 0$ such that $J(u) = 0$. The infimum, of course, is zero.

This illustrates the basic fact that the specification of the class of admissible functions is crucial in determining whether a minimizing function exists.

EXERCISE

1. Does $J(u)$ possess a minimum in the class of functions determined by $u' \in L^2(0, T)$, $u(0) = c \neq 0$?

4.8. Nonexistence of a Minimum—II

Consider next a more "legitimate" example. Let

$$J(u) = \int_0^{4\pi} (u'^2 - u^2) \, dt, \qquad u(0) = 0. \tag{4.8.1}$$

The Euler equation is readily seen to be

$$u'' + u = 0,$$
$$u(0) = 0, \qquad u'(4\pi) = 0. \tag{4.8.2}$$

This possesses the unique solution $u = 0$. However, the minimum value of $J(u)$ is *not* zero, as we shall show.

Consider the trial function

$$u = k \sin t/2. \tag{4.8.3}$$

We see that

$$J(k \sin t/2) = k^2 \left[\int_0^{2\pi} \left[\frac{\cos^2 t/2}{4} - \sin^2 t/2 \right] dt \right.$$

$$= -\frac{3k^2}{4} \int_0^{4\pi} \sin^2 t/2 \, dt. \tag{4.8.4}$$

As k increases without bound, $J(k \sin t/2)$ assumes arbitrarily large negative values. Hence, no minimum exists, despite the fact that the Euler equation possesses a unique solution.

On the other hand, it is easy to show that the functional

$$J_T(u) = \int_0^T (u'^2 - u^2) \, dt \tag{4.8.5}$$

possesses an absolute minimum of zero in the class of functions defined by $u(0) = 0$, $u' \in L^2(0, T)$, provided that $T \ll 1$. We shall occasionally use the notation $T \ll 1$ to express the condition that T is sufficiently small. Sturm-Liouville theory yields the precise bound on T; see Volume I.

EXERCISE

1.   What is the precise bound on T?

4.9.   Nonexistence of a Minimum—III

Finally, consider the functional

$$J(u) = \int_0^T (u'^2 - u^4) \, dt. \tag{4.9.1}$$

We assert that J possesses no minimum in the class of functions defined by $u(0) = c \neq 0$, $u' \in L^2(0, T)$, no matter how small T is.

To see this, consider the trial function $u = ce^{kt}$. Then

$$J(ce^{kt}) = \int_0^T [c^2 k^2 e^{2kt} - c^4 e^{4kt}] \, dt. \tag{4.9.2}$$

It is clear that no matter how small c and T are, we can choose k large enough to give $J(ce^{kt})$ an arbitrarily large negative value. Hence, J possesses no absolute minimum. Nonetheless, as we shall see below, the associated Euler equation possesses at least one solution for small T.

From what has preceded, we can conclude that the formalism of the calculus of variations must be applied with care and forethought.

4.10. Existence of Solution of Euler Equation

As a first step, we wish to demonstrate that the equation

$$u'' - g(u) = 0,$$
$$u(0) = c, \qquad u'(T) = 0,$$

(4.10.1)

possesses *a* solution under suitable assumptions concerning c, T, and $g(u)$. Integrating once from t to T, and using the condition $u'(T) = 0$, we have

$$u'(t) = -\int_t^T g(u) \, dt_1,$$

(4.10.2)

whence a further integration from 0 to t yields

$$u(t) = c - \int_0^t \left[\int_{t_2}^T g(u) \, dt_1 \right] dt_2.$$

(4.10.3)

Thus the original nonlinear differential equation has been converted into a nonlinear integral equation. One of the advantages of an integral equation formulation is that the boundary conditions are automatically incorporated into the equation.

Write this in the form

$$u = F(u).$$

(4.10.4)

The solution of the original equation is thus a *fixed point* of the transformation F. Conversely, every fixed point of F is a solution of (4.10.1), since every solution of (4.10.3) satisfies (4.10.1). We can, therefore, use the fixed-point theory of Birkhoff-Kellogg to establish the existence of a solution of the Euler equation under various sets of hypotheses. We shall not, however, follow this route since it is nonconstructive and not needed at the moment. In more complex situations, however, fixed-point theory is an essential tool, and indeed the only available tool. To develop this properly, the more modern and sophisticated results of Schauder-Leray and others are required.

4.11. Successive Approximations

We will employ the method of successive approximations to establish the existence of a solution of (4.10.4) and thus of (4.10.1). Let $u_0 = c$ be the initial guess and let the sequence $\{u_n\}$ be defined inductively for $n \geq 1$ by means of the relation

$$u_n = F(u_{n-1}).$$

(4.11.1)

Our aim is to show that $\{u_n\}$ converges uniformly to a solution of the Euler equation provided that $g(u)$ possesses a bounded derivative in the neighborhood of $u = c$ and that $T \ll 1$. We will obtain precise, but quite pessimistic, bounds for the allowable magnitude of T.

Let us begin with the observation that integration by parts of (4.10.3) yields

$$F(u) = c + t \int_t^T g(u) \, dt_2 + \int_0^t t_2 \, g(u) \, dt_2$$

$$= c + \int_0^T k(t, t_2, T) g(u) \, dt_2, \qquad (4.11.2)$$

where the kernel, the Green's function, is defined by

$$k(t, t_2, T) = t_2, \quad 0 \le t_2 \le t \le T,$$

$$= t, \quad t \le t_2 \le T. \qquad (4.11.3)$$

Hence, we have the simple bound

$$|k(t, t_2, T)| \le T, \quad 0 \le t, t_2 \le T. \qquad (4.11.4)$$

Let us introduce the constants, dependent on T,

$$b_1 = \max_{|u-c| \le T} |g(u)|,$$

$$b_2 = \max_{|u-c| \le T} |g'(u)|. \qquad (4.11.5)$$

Let us show inductively that

$$|u_n(t) - c| \le T \qquad (4.11.6)$$

for $0 \le t \le T$, provided that T is sufficiently small. As we shall see, a precise estimate is $b_1 T \le 1$. The result is certainly true for $n = 0$ since $u_0 = c$. For $n = 1$, we have, for $0 \le t \le T$,

$$|u_1(t) - c| \le \int_0^T |k(t, t_1, T) g(u_0)| \, dt_1$$

$$\le T \int_0^T |g(u_0)| \, dt_1 \qquad (4.11.7)$$

$$\le T(b_1 T) \le T,$$

provided that $b_1 T \le 1$. Inductively, we see using the same argument that $|u_n(t) - c| \le T$ for $0 \le t \le T$ implies the same inequality for $n + 1$, provided that $b_1 T \le 1$.

It remains to establish convergence of the sequence $\{u_n\}$. We have

$$u_{n+1}(t) - u_n(t) = \int_0^T k(t, t_1, T)[g(u_n) - g(u_{n-1})] \, dt_1. \qquad (4.11.8)$$

Hence,

$$|u_{n+1}(t) - u_n(t)| \leq T \int_0^T |g(u_n) - g(u_{n-1})| \, dt_1$$

(4.11.9)

$$\leq Tb_2 \int_0^T |u_n - u_{n-1}| \, dt_1.$$

We have used the fact that $|g(v) - g(w)| \leq b_2 |v - w|$ for both v and w in the interval $|u - c| \leq T$. Here b_2 is a bound on $g'(u)$. Thus,

$$\max_t |u_{n+1}(t) - u_n(t)| \leq T^2 b_2 \left(\max_t |u_n - u_{n-1}| \right).$$

(4.11.10)

Hence, if $T^2 b_2 < 1$, the series

$$\sum_n \max_t |u_{n+1} - u_n|$$

(4.11.11)

converges. Thus, the series $\sum (u_{n+1} - u_n)$ converges uniformly for $0 \leq t \leq T$, and consequently, as $n \to \infty$, u_n converges uniformly to a continuous function $u(t)$ for $0 \leq t \leq T$. This function satisfies the integral equation

$$u(t) = c + \int_0^T k(t, t_1, T)g(u) \, dt_1,$$

(4.11.12)

since uniform convergence permits us to pass to the limit under the integral sign as $n \to \infty$.

Referring to the explicit form of $k(t, t_1, T)$, we see that u is differentiable and satisfies the original differential equation (4.10.1).

Hence, if $g(u)$ possesses a bounded derivative in some neighborhood of c, say $|u - c| \leq c_0$, we can conclude that the Euler equation has a solution for sufficiently small T.

EXERCISES

1. Similarly establish the existence of a solution to $u'' - g(u) = 0$, $u(0) = c_0$, $u(T) = c_2$.
2. Consider also the case where the boundary conditions are $u(0) = c_1$, $u(T) + bu'(T) = c_2$.
3. Is the limit of the solution of Problem 2 as $b \to 0$ a solution of Problem 1?
4. Consider the case where $u'' - g(u) = 0$, $u(0) = c_1$, $k(u(T), u'(T)) = 0$, under suitable conditions on k.
5. Is it essential that g possess a continuous derivative?
6. Obtain a specific bound for T in the case $u'' - u^{2k+1} = 0$, $u(0) = 1$, $u'(T) = 0$.
7. Use the solution obtained from the first integral, $u'^2 - (2u^{2k+2}/(2k + 2)) = c_3$, to obtain the best possible results.

4.12. Conditional Uniqueness

Suppose that $v(t)$ is another solution of (4.10.1). Then

$$u - v = \int_0^T k(t, t_1, T)[g(u) - g(v)] \, dt_1. \qquad (4.12.1)$$

Let $\max_{0 \leq t \leq T} |v| = b_3$ and

$$|g(u) - g(v)| \leq b_4 |u - v| \qquad (4.12.2)$$

for $|u - c| \leq T$, $|v| \leq b_3$. Then

$$|u - v| \leq T \int_0^T |g(u) - g(v)| \, dt_1 \qquad (4.12.3)$$

$$\leq Tb_4 \int_0^T |u - v| \, dt_1.$$

Hence,

$$\max_{0 \leq t \leq T} |u - v| \leq T^2 b_4 \left(\max_{0 \leq t \leq T} |u - v| \right). \qquad (4.12.4)$$

Thus, if $T^2 b_4 < 1$, we see that

$$\max_{0 \leq t \leq T} |u - v| = 0, \qquad (4.12.5)$$

whence $u \equiv v$ in $[0, T]$.

Thus, we have *conditional* uniqueness. The solution is unique within a class of functions which are suitably bounded over $[0, T]$. This is not surprising as the example of the next section shows.

4.13. A Plethora of Geodesics

An easy way of seeing that there can be variational problems with Euler equations which possess many solutions, indeed denumerably many solutions, is to consider the problem of determining a geodesic between two points on a torus, as in Figure 4.1. If the points P and Q are close enough together, the

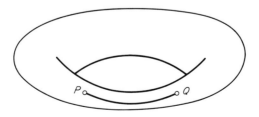

Figure   4.1

geodesic connecting them will be as shown. However, there is also a relative minimum distance attained within the class of all curves connecting P and Q which loop around the torus once; see Figure 4.2. Similarly, we have a minimum in the class of curves looping twice, and so on.

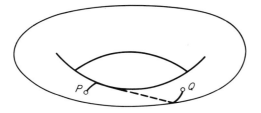

Figure 4.2

All of these minima, stationary points of the functional, are furnished as solutions of the Euler equation with the same boundary conditions, namely the curves must pass through P and Q.

EXERCISES

1. Determine the Euler equation for the foregoing problem.
2. Examine similarly the determination of geodesics on a sphere with a hill.
3. Consider the minimum distance between two points on a sphere to exhibit an Euler equation possessing two solutions, one of which is an absolute minimum. What is the other?

4.14. A Priori Bounds

In some cases, we can readily determine a priori bounds on the desired solution of the Euler equation and thereby obtain far more satisfactory results. Consider, for example, the problem of minimizing

$$J(u) = \int_0^T (u'^2 + 2h(u))\, dt, \tag{4.14.1}$$

where we suppose that $u(0) = c$ and that $h(u) \geq 0$ for all u. This is a natural condition in control theory if we think of $h(u)$ as a cost. Since $u = c$ is an acceptable trial function, we must have the inequality

$$\int_0^T u'^2\, dt \leq \int_0^T (u'^2 + 2h(u))\, dt \leq 2 \int_0^T h(c)\, dt = 2h(c)T \tag{4.14.2}$$

for a minimizing function. Thus, we are interested only in functions which satisfy the bound

$$\int_0^T u'^2 \, dt \leq 2h(c)T. \tag{4.14.3}$$

Since

$$u(t) - c = \int_0^t u' \, dt_1, \tag{4.14.4}$$

we have, using the Cauchy-Schwarz inequality,

$$(u(t) - c)^2 = \left(\int_0^t u' \, dt_1\right)^2 \leq t \int_0^t u'^2 \, dt_1$$

$$\leq T \int_0^T u'^2 \, dt_1 \leq 2h(c)T^2. \tag{4.14.5}$$

Thus, we can safely limit our attention to functions satisfying the constraint

$$\max_{0 \leq t \leq T} |u(t) - c| \leq T(2h(c))^{1/2}. \tag{4.14.6}$$

The foregoing argument in Section 4.12 shows that for small T there is exactly one solution of the Euler equation satisfying this constraint. This solution may be obtained using successive approximations as indicated. In the next chapter we will indicate a far more adroit utilization of the method of successive approximations, which is an offshoot of the general theory of quasilinearization.

EXERCISES

1. Show that $u'' + u = 0$, $u(0) = c$, $u'(T) = 0$ has a unique solution if $T \ll 1$. Is this condition necessary?
2. Does $u'' + u + u^3 = 0$, $u(0) = c$, $u'(T) = 0$ possess a unique solution for all c and T?

4.15. Convexity

To obtain more satisfying results we must impose some stronger condition on $g(u)$. A natural one is that of *convexity*. Geometrically, convexity asserts that the curve $v = g(u)$ always lies below its chord, as in Figure 4.3.

We shall take an analytic version of this,

$$g''(u) \geq 0, \tag{4.15.1}$$

and consider only the case of strict convexity, $g''(u) > 0$.

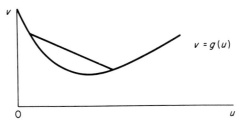

Figure 4.3

EXERCISES

1. If $g(u)$ is strictly convex, show that $g(\lambda u + (1 - \lambda)v) \leq \lambda g(u) + (1 - \lambda)g(v)$ for $0 \leq \lambda \leq 1$.
2. If $g(u)$ is strictly convex, show that $g(u) = \max_v [g(v) + (u - v)g'(v)]$.
3. What is a geometric interpretation of this analytic relation?
4. Show that strict convexity is equivalent to the condition that any three points on the curve are the vertices of a triangle with positive area.

4.16. Sufficient Condition for Absolute Minimum

Let us now suppose that $h(u)$ is convex and that we have in some fashion established the existence of a solution of

$$u'' - h'(u) = 0, \qquad u(0) = c, \qquad u'(T) = 0. \tag{4.16.1}$$

Then we assert that this solution furnishes the absolute minimum of

$$J(u) = \int_0^T (u'^2 + 2h(u)) \, dt \tag{4.16.2}$$

over the class of functions satisfying the conditions $u' \in L^2(0, T)$, $u(0) = c$.

The proof is direct. We have, for any function v such that $v(0) = 0$, $v' \in L^2(0, T)$,

$$J(u + v) = \int_0^T [(u' + v')^2 + 2h(u + v)] dt$$

$$= J(u) + 2 \int_0^T (u'v' + h'(u)v) \, dt \tag{4.16.3}$$

$$+ \int_0^T (v'^2 + h''(\theta)v^2) \, dt.$$

Here $\theta(t)$ is a function of t lying on the interval $[u, u + v]$ for $0 \leq t \leq T$.

We are using the standard mean-value theorem of calculus for the expansion of $h(u + v)$, namely

$$h(u + v) = h(u) + vh'(u) + \frac{v^2}{2} h''(\theta). \qquad (4.16.4)$$

The second term in (4.16.3) vanishes upon an integration by parts,

$$\int_0^T (u'v' + h'(u)v) \, dt = u'v \Big]_0^T + \int_0^T (-u'' + h'(u))v \, dt = 0. \qquad (4.16.5)$$

Hence,

$$J(u + v) = J(u) + \int_0^T (v'^2 + h''(0)v^2) \, dt$$

$$\geq J(u) + \int_0^T v'^2 \, dt. \qquad (4.16.6)$$

Hence, $J(u + v) > J(u)$ if v', and hence, v is not identically zero.

EXERCISES

1. For fixed T show that the condition $h'' \geq 0$ can be replaced by $h'' \geq -k_1(T)$. Determine $k_1(T)$.
2. Show that we can avoid the mean-value theorem, using only the relation $h(u + v) \geq h(u) + vh'(u)$, valid for a convex function.

4.17. Uniqueness of Solutions of Euler Equation

The fact that any solution of (4.16.1) furnishes the absolute minimum of $J(u)$ in the case where $h(u)$ is convex enables us to assert the uniqueness of the solution of (4.16.1) in this case. Thus, we determine uniqueness of the solution of the differential equation by considering the behavior of an auxiliary functional, in this case the control functional, the criterion function. We employed this technique extensively in Volume I.

4.18. Demonstration of Minimizing Property—Small T

If we do not impose a strong condition such as convexity, the best we can hope to do is establish a restricted minimizing property, local in function space and local in time. We cannot hope to do better than we can in ordinary calculus.

Returning to (4.16.6), we have

$$J(u + v) = J(u) + \int_0^T (v'^2 + h''(\theta)v^2) \, dt. \qquad (4.18.1)$$

Let v be such that $|h''(\theta)| \le k_1$. Then

$$J(u + v) > J(u) + \int_0^T (v'^2 - k_1 v^2) \, dt. \qquad (4.18.2)$$

We know that the quadratic functional is positive definite for all nontrivial v such that $v(0) = v'(T) = 0$, provided that $T \ll 1$. The precise bound on T depends on k_1. Hence, we have the desired local minimization.

4.19. Discussion

The foregoing analysis exhibits very clearly some of the basic strengths and weaknesses of the calculus of variations in handling control processes. The theory provides a powerful constructive tool for treating control processes, the Euler equation. In return, we must restrict our attention to small time intervals and to local variations in the time history of the system.

Control theory, on the other hand, is primarily interested in long-term processes, and in absolute minima, which is to say the most efficient operation. This means that more powerful and versatile methods are required. As we shall subsequently see, by combining the calculus of variations with other approaches we can in many cases determine optimal control. It is to be expected that no single method will suffice in the treatment of control processes of major significance, as we have already noted.

4.20. Solution as Function of Initial State

Let us turn to the equation

$$u'' - g(u) = 0,$$
$$u(0) = c, \qquad u'(T) = 0, \qquad (4.20.1)$$

(replacing $h'(u)$ by $g(u)$), and derive some results of independent interest concerning the nature of the solution. These will be needed in Chapter 6.

It is clear that the solution depends on the initial value c and the duration of the process T. Hence, we may regard the solution at time t, $0 \le t \le T$, as a function of the initial state c and the duration T. Write

$$u \equiv u(t) \equiv u(t, c, T). \qquad (4.20.2)$$

Our aim is to study u as a function not only of t but also of c and T. In particular, we wish to obtain some relations connecting u_c and u_T which we will use to determine the dependence of the minimum value of $J(u)$ and of the missing initial condition $u'(0)$ on c and T.

To begin with, we wish to demonstrate that u_c exists in some region of (c, T)-space. A simple way to do this is to utilize the original recurrence relation,

$$u_n'' - g(u_{n-1}) = 0,$$
$$u_n(0) = c, \qquad u_n'(T) = 0. \tag{4.20.3}$$

One of the principal advantages of the method of successive approximations is that it affords a convenient bootstrap.

Let us assume that $g(u)$ and $g'(u)$ exist in some u-interval, say $|u| \leq c_0$. Specifically, let $|g'(u)|$ be uniformly bounded in this interval by a_1,

$$|g'(u)| \leq a_1. \tag{4.20.4}$$

Then, returning to the proof of the existence of a solution in Section 4.11, we know that the sequence $\{u_n\}$ satisfies the bound

$$|u_n - c| \leq T \tag{4.20.5}$$

for $|c| \leq c_0$, provided that $T \ll 1$.

Turning to (4.20.3), we see that $(u_n)_c$ exists for all $u \geq 0$. We have

$$(u_n)_c'' - g'(u_{n-1})(u_{n-1})_c = 0,$$
$$u_n(0)_c = 1, \qquad u_n'(T)_c = 0. \tag{4.20.6}$$

This may be written so as to be a relation for the sequence $\{(u_n)\}_c$, namely, writing $v_n = (u_n)_c$, we have

$$v_n'' - g'(u_{n-1})v_{n-1} = 0,$$
$$v_n(0) = 1, \qquad v_n'(T) = 0. \tag{4.20.7}$$

It follows, using the argument in Section 4.11, that if we restrict c to a subinterval, say $|c| \leq c_0/2$, and take $T \ll 1$, then the sequence $\{u_n\}_c$ converges uniformly for $0 \leq t \leq T$. If

$$(u_n)_c \to v,$$
$$(u_n) \to u, \tag{4.20.8}$$

as $n \to \infty$, it follows from the relation

$$u_n = \int (u_n)_c \, dc \qquad (4.20.9)$$

that $v = u_c$. Similarly, we can establish the existence of higher partial derivatives, under the assumption that g possesses the requisite derivatives.

Hence, we can assert that there is a region in (c, T)-space, $|c| \le c_0/2$, $0 \le T \le T_0$, with the property that $u(t, c, T)$ exists for $0 \le t \le T$, is a conditionally unique solution of the Euler equation and possesses a partial derivative with respect to c.

EXERCISES

1. If $g(u)$ is analytic in a region $|u| \le u_0$, show that $u(t, c, T)$ is analytic in c in a suitable c-region.
2. Can we start from the Euler equation $u'' - g(u) = 0$, $u(0) = c$, $u'(T) = 0$ and show directly that u_c exists and satisfies the *linear* equation $w'' - g'(u)w = 0$, $w(0) = 1$, $w'(T) = 0$?
3. Obtain corresponding results concerning the dependence of $u'(0)$ and $u(T)$ upon c.

4.21.   Solution as Function of Duration of Process

Let us now consider the dependence on T. It is not immediate from the differential equation how we go about computing the partial derivative $(u_n)_T$. We can, however, use the integral relation of Section 4.11 to obtain the desired result. We have

$$u_n = F(u_{n-1}) = c + t \int_t^T g(u_{n-1}) \, dt_2 + \int_0^t t_2 \, g(u_{n-1}) \, dt_2, \qquad (4.21.1)$$

for $0 \le t \le T$. It follows, for $t \le T$, that

$$(u_n)_T = tg(u_{n-1}(T)) + t \int_t^T g'(u_{n-1})(u_{n-1})_T \, dt_2$$

$$+ \int_0^{t_2} g'(u_{n-1})(u_{n-1})_T \, dt_2$$

$$= tg(u_{n-1}(T)) + \int_0^T k(t, t_2, T)g'(u_{n-1})(u_{n-1})_T \, dt_2. \qquad (4.21.2)$$

As above, we readily establish the convergence of $\{(u_n)_T\}$ in a region $|c| \le c_0/2$, $0 \le T \le T_1$, where T_1 may be less than T_0, and again, that the limit of $(u_n)_T$ must be u_T.

4.22.   The Return Function

Let us deduce some consequences of the fact that the minimum value of $J(u)$ may also be regarded as a function of c and T. Introduce the function $f(c, T)$ defined by

$$f(c, T) = \min_{u} \int_{0}^{T} (u'^2 + 2h(u))\, dt. \tag{4.22.1}$$

We call this the *return function* as in the discrete case. To ensure that the function is well defined, we can impose the strong condition that $h(u)$ is convex.

This definition is then equivalent to the statement

$$f(c, T) = \int_{0}^{T} (u'^2 + 2h(u))\, dt, \tag{4.22.2}$$

where u is determined by

$$u'' - h'(u) = 0,$$
$$u(0) = c, \qquad u'(T) = 0. \tag{4.22.3}$$

Let us compute some partial derivatives. We have

$$\begin{aligned}
f_c &= \int_{0}^{T} [2u'u'_c + 2h'(u)u_c]\, dt \\
&= 2u'u_c \Big]_{0}^{T} + 2\int_{0}^{T} [-u'' + h'(u)]u_c\, dt \\
&= -2u'(0),
\end{aligned} \tag{4.22.4}$$

upon taking account of the Euler equation and the various boundary conditions at $t = 0$ and T.

Similarly,

$$\begin{aligned}
f_T &= [u'(T)^2 + 2h(u(T))] + \int_{0}^{T} [2u'u'_T + 2h'(u)u_T]\, dt \\
&= 2h(u(T)) + 2u'u_T \Big]_{0}^{T} + 2\int_{0}^{T} [-u'' + h'(u)]\, u_T\, dt \\
&= 2h(u(T)).
\end{aligned} \tag{4.22.5}$$

4.23.   A Nonlinear Partial Differential Equation

Returning to the Euler equation associated with

$$J(u) = \int_{0}^{T} (u'^2 + 2h(u))\, dt, \tag{4.23.1}$$

namely

$$u'' - h'(u) = 0, \qquad (4.23.2)$$

we have

$$\int_0^T u'(u'' - h'(u)) \, dt_1 = 0. \qquad (4.23.3)$$

Hence,

$$u'(T)^2 - u'(0)^2 = 2h(u(T)) - 2h(u(0)), \qquad (4.23.4)$$

the well known first integral. Since $u(0) = c$, $u'(T) = 0$, this becomes

$$-u'(0)^2 = 2h(u(T)) - 2h(c). \qquad (4.23.5)$$

Referring to the results of (4.22.4) and (4.22.5), we see that (4.23.5) may be written

$$f_T = 2h(c) - f_c^2/4. \qquad (4.23.6)$$

This is a nonlinear partial differential equation with the initial condition

$$f(c, 0) = 0. \qquad (4.23.7)$$

4.24. Discussion

The foregoing derivation is noticeably ad hoc. It is rather clearly a validation of something we know from other considerations to be true, rather than a derivation of results from first principles. In the chapter on dynamic programming and the calculus of variations, Chapter Six, we shall show how this equation arises in a natural and direct fashion.

Relations of the type appearing in (4.23.6) are part of the classical theory of Hamilton and Jacobi for dynamical systems.

Once this equation (4.23.6), has been established, the question arises of bypassing the Euler equation and its difficulties and focusing directly upon the partial differential equation. As we shall see, there are considerable advantages to retaining the control theory framework. Indeed, in this way an entirely new approach to many classes of partial differential equations is opened up.

EXERCISES

1. Consider the case where $J(u) = \int_0^T (u'^2 + u^2) \, dt$ and where we wish to minimize over functions such that $u(0) = c$, $u' \in L^2(0, T)$. Show directly, without use of the Euler equation, that $f(c, T) = \min_u J(u) = r(T)c^2$. (*Hint*: Consider the change of variable $u = cv$.)

2.   Hence, using the partial differential equation, show that $r(T)$ satisfies a Riccati differential equation $r' = 1 - r^2$, $r(0) = 0$.

3.   Consider the case where $J(u) = \int_0^T (u'^4 + u^4)\, dt$. Show that $f(c, T) = c^4 r_4(T)$ and find the ordinary differential equation for $r_4(T)$.

4.   Consider the case where $J(u) = \int_0^T (u'^{2n} + u^{2n})\, dt$. Write $f(c, T) = c^{2n} r_{2n}(T)$ and obtain the equation for $r_{2n}(T)$. Does the equation have a limiting form as $n \to \infty$? Does $r_{2n}(T)^{1/2n}$ possess a limit as $n \to \infty$? Does this limit function satisfy a limiting differential equation? Is there a corresponding limiting control process?

5.   Show that the nonlinear partial differential equation in (4.23.6) may be written

$$f_T = \min_v\ [c^2 + v^2 + vf_c].$$

6.   Write $\varphi(c, T) = \min_u [\int_0^T g(u, u')\, dt]$. Obtain formally an equation for φ corresponding to (4.23.6) and show that it may be written

$$\phi_T = \min_v\ [g(c, v) + v\phi_c].$$

7.   What conditions on $g(u, u')$ are sufficient to provide a rigorous basis for this result?

4.25.   More General Control Processes

Let us next consider some control processes of a more general type, beginning with the one-dimensional case. We wish to minimize

$$J(u, v) = \int_0^T h(u, v)\, dt, \tag{4.25.1}$$

where u and v are related by the equation

$$u' = g(u, v), \qquad u(0) = c. \tag{4.25.2}$$

If we wish, we can make a change of variable

$$w = g(u, v), \tag{4.25.3}$$

whence $v = k(u, w)$, so that the problem becomes: Minimize

$$J_1(u) = \int_0^T h(u, k(u, u'))\, dt = \int_0^T h_1(u, u')\, dt. \tag{4.25.4}$$

This is the type of problem we have already treated. However, we do not pursue this path because the transformation is almost always an unnatural one which conceals the important variables. Let us then consider the original problem.

We shall begin with a formal derivation of the Euler equation without making precise the domains of the state and control variable. Let u, v be a minimizing pair, and consider $J(u + \varepsilon\bar{u}, v + \varepsilon\bar{v})$, where ε is a scalar and $\bar{u}, \bar{v}$ are "arbitrary." Then, proceeding as before,

$$J(u + \varepsilon\bar{u}, v + \varepsilon\bar{v}) = \int_0^T h(u + \varepsilon\bar{u}, v + \varepsilon\bar{v}) \, dt$$

$$= J(u, v) + \varepsilon \int_0^T (\bar{u}h_u + \bar{v}h_v) \, dt + 0(\varepsilon^2), \qquad (4.25.5)$$

and

$$u' + \varepsilon\bar{u}' = g(u + \varepsilon\bar{u}, v + \varepsilon\bar{v}) = g(u, v) + \varepsilon(\bar{u}g_u + \bar{v}g_v) + 0(\varepsilon^2). \qquad (4.25.6)$$

Hence, $\bar{u}$ and $\bar{v}$ are related by the equation

$$\bar{u}' = \bar{u}g_u + \bar{v}g_v, \qquad \bar{u}(0) = 0. \qquad (4.25.7)$$

From (4.25.5) we have the variational condition

$$\int_0^T (\bar{u}h_u + \bar{v}h_v) \, dt = 0. \qquad (4.25.8)$$

To eliminate one of the two functions, $\bar{u}, \bar{v}$, we can proceed as follows. Solve for $\bar{v}$ in (4.25.7)

$$\bar{v} = (\bar{u}' - \bar{u}g_u)/g_v, \qquad (4.25.9)$$

and substitute in (4.25.8)

$$\int_0^T \left(\bar{u}h_u + \frac{h_v}{g_v}(\bar{u}' - \bar{u}g_u) \right) dt = 0,$$

$$\int_0^T \left[\frac{h_v\bar{u}'}{g_v} + \bar{u}\left(h_u - \frac{g_u h_v}{g_v} \right) \right] dt = 0. \qquad (4.25.10)$$

Integration by parts now yields in familiar fashion the Euler equation

$$\frac{d}{dt}\left(\frac{h_v}{g_v} \right) - \left(\frac{h_u g_v - h_v g_u}{g_v} \right) = 0. \qquad (4.25.11)$$

An immediate objection to this procedure is the fact that we have no guarantee g_v is nonzero throughout $[0, T]$. Note that g_v disappears from the denominator in (4.25.11) after differentiation. An alternate route which is a bit longer, but guaranteed, is the following: Let us solve (4.25.7) for $\bar{u}$. We have

$$\bar{u} = \exp\left(\int_0^t g_u \, dt_1 \right)\left[\int_0^t \exp -\left(\int_0^{t_2} g_u \, dt_1 \right) g_v \bar{v} \, dt_2 \right]. \qquad (4.25.12)$$

Substituting in (4.25.8), the result is

$$\int_0^T \left[h_u \exp \left(\int_0^t g_u \, dt_1 \right) [\cdots] + \bar{v} h_v \right] dt = 0. \qquad (4.25.13)$$

Integrating by parts, the first term yields

$$\left\{ \int_t^T h_u \exp \left(\int_0^{t_2} g_u \, dt_1 \right) dt_2 \right\} [\cdots] \Big|_0^T + \int_0^T \left[\exp - \left(\int_0^{t_2} g_u \, dt_1 \right) g_v \bar{v} \right.$$

$$\left. \times \int_{t_2}^T h_u \exp \left(\int_0^t g_u \, dt_1 \right) dt_2 + \bar{v} h_v \right] dt = 0$$

$$(4.25.14)$$

The integrated terms drop out and the variational equation is thus

$$\exp - \left(\int_0^t g_u \, dt_1 \right) g_v \int_t^T h_u \exp \left(\int_0^t g_u \, dt_1 \right) dt + h_v = 0. \qquad (4.25.15)$$

We leave it to the reader to show that this is equivalent to (4.25.11) if $g_v \neq 0$.

4.26. Discussion

There is little difficulty in deriving various local existence and uniqueness theorems for the Euler equation using the method of successive approximations, and various conditional minimization results for $J(u, v)$. It is essential that the student understand how to obtain these results if they are desired in any particular case. However, since they are not very satisfying results at best, we shall not dwell upon them or devote any further time or effort to describing them. References to extensive discussions of these matters will be found at the end of this chapter.

EXERCISES

1. If the original equation connecting u and v were (4.25.11) rather than (4.25.2), would we obtain $u' = g(u, v)$ as the Euler equation?
2. Obtain (4.25.10) by using the criterion function of (4.25.4), assuming all transformations are permissible.

4.27. Multidimensional Control Processes—I

Although the theory of multidimensional control processes parallels that of one-dimensional control processes to a great extent, there are a few analytic differences, due to dimension, that are worth examining. The major difference, however, enters in connection with the basic problem of effective computational solution. We have discussed this at some length in Volume I, and we shall return to it in the next chapter.

The first multidimensional case of importance involves the minimization of the functional

$$J(x) = \int_0^T [(x', x') + 2g(x)] \, dt. \tag{4.27.1}$$

Let x belong to the class of functions for which $x(0) = c$ and $x' \in L^2(0, T)$.

We leave it to the reader to show that the Euler equation takes the form

$$x'' - \text{grad } g = 0,$$
$$x(0) = c, \qquad x'(T) = 0, \tag{4.27.2}$$

where

$$\text{grad } g = \begin{pmatrix} \dfrac{\partial g}{\partial x_1} \\ \vdots \\ \dfrac{\partial g}{\partial x_N} \end{pmatrix}. \tag{4.27.3}$$

EXERCISE

1.   What is the Euler equation associated with $J(x) = \int_0^T [(x', x') + (x, Ax)] \, dt$?

4.28.   Auxiliary Results

To obtain analogues for the scalar case we need some results concerning multidimensional convexity. If $g(x_1, x_2, \ldots, x_N) = g(x)$ is a scalar function with the property that the matrix

$$H = \left(\frac{\partial^2 g}{\partial x_i \, \partial x_j} \right), \qquad i, j = 1, 2, \ldots, N, \tag{4.28.1}$$

is positive definite for all x, we say that it is strictly convex. As in the scalar case we can show directly that

$$g(x) = \max_y \, [g(y) + (x - y, \text{grad } g)], \tag{4.28.2}$$

if $g(x)$ is strictly convex. It follows that

$$g(x) \geq g(y) + (x - y, \text{grad } g), \tag{4.28.3}$$

for all y, an inequality we can employ in the same way the one-dimensional result was used in Section 4.16.

EXERCISES

1. Establish (4.28.2).
2. Consider the function of the scalar variable, defined by $f(s) = g(sx + (1 - s)y)$, $0 \le s \le 1$. Apply the mean-value theorem to $f(s)$ to obtain a corresponding mean-value theorem for $g(x)$. (*Hint*: Write $f(s) = f(0) + sf'(0) + s^2 f''(\theta)/2$, and calculate what $f(0), f'(0), f''(\theta)$ are.)
3. If $g(x)$ is strictly convex, show that $g(sx + (1 - s)y) \le sg(x) + (1 - s)g(y)$ for $0 \le s \le 1$.
4. What is the geometric significance of this result?

4.29. Multidimensional Control Processes—II

Let us now discuss the problem of minimizing the functional

$$J(x, y) = \int_0^T h(x, y) \, dt, \tag{4.29.1}$$

where x and y are related by the equation

$$x' = g(x, y), \qquad x(0) = c. \tag{4.29.2}$$

Our aim is to derive the Euler equation. Although the basic idea, of course, remains the same, there are some points arising from the multidimensional nature of the problem which are worth examining in detail. Proceeding as before, we write

$$J(x + \varepsilon \bar{x}, y + \varepsilon \bar{y}) = \int_0^T h(x + \varepsilon \bar{x}, y + \varepsilon \bar{y}) \, dt$$

$$= \int_0^T h(x, y) \, dt + \varepsilon \int_0^T [(h_x, \bar{x}) + (h_y, \bar{y})] \, dt + 0(\varepsilon^2). \tag{4.29.3}$$

Here h_x is the vector whose components are $\partial h/\partial x_i$, $i = 1, 2, \ldots, N$, and h_y is determined similarly. These are partial gradients. Hence, the variational condition is

$$\int_0^T [(h_x, \bar{x}) + (h_y, \bar{y})] \, dt = 0. \tag{4.29.4}$$

Similarly, the differential equation connecting x and y yields the variational equation

$$\bar{x}' = J_x(g)\bar{x} + J_y(g)\bar{y}, \qquad \bar{x}(0) = 0, \tag{4.29.5}$$

where $J_x(g)$ and $J_y(g)$ are the Jacobian matrices,

$$J_x(g) = \left(\frac{\partial g_i}{\partial x_j}\right), \qquad J_y(g) = \left(\frac{\partial g_i}{\partial y_j}\right). \tag{4.29.6}$$

The solution of (4.29.5) may be written

$$\bar{x} = X(t) \int_0^t X(s)^{-1} J_y(g) \bar{y} \, ds, \tag{4.29.7}$$

where X is the solution of the matrix equation

$$X' = J_x(g)X, \qquad X(0) = I. \tag{4.29.8}$$

The variational equation in (4.29.4) then takes the form

$$\int_0^T \left[\left(h_x, X(t) \int_0^t X(s)^{-1} J_y(g) \bar{y} \, ds \right) + (h_y, \bar{y}) \right] dt = 0. \tag{4.29.9}$$

Integrating by parts then yields as before the equation

$$\int_0^T \left[\left(\int_t^T X(t_1)^* h_x \, dt_1, X(t)^{-1} J_y(g) \bar{y} \right) + (h_y, \bar{y}) \right] dt = 0 \tag{4.29.10}$$

(here X^* is the transpose of X), whence

$$J_y(g)^*(X(t)^*)^{-1} \int_t^T X(t_1)^* h_x \, dt_1 + h_y = 0. \tag{4.29.11}$$

If $J_y(g)^*$ is nonsingular, we can write this in the form

$$\int_t^T X(t_1)^* h_x \, dt_1 + X(t)^*(J_y(g)^*)^{-1} h_y = 0, \tag{4.29.12}$$

and eliminate $X(t_1)^*$ by differentiation. Since $X(t)$ satisfies (4.29.8), we know that $X(t)^*$ satisfies the equation

$$(X^*)' = -X^* J_x(g). \tag{4.29.13}$$

Thus, differentiating (4.29.12), we obtain

$$-X(t)^* h_x - X(t)^* J_x(g)(J_y(g)^*)^{-1} h_y + X(t)^*[\{J_y(g)^*\}^{-1} h_y]' = 0, \tag{4.29.14}$$

or, finally, the desired equation,

$$-h_x - J_x(g)(J_y(g)^*)^{-1} h_y + [\{J_y(g)^*\}^{-1} h_y]' = 0. \tag{4.29.15}$$

The additional boundary condition comes from (4.29.11),

$$h_y \Big|_{t=T} = 0. \tag{4.29.16}$$

EXERCISE

1. Obtain the Euler equation for the minimization of $J(x, y) = \int_0^T [(x, x) + (y, y)] \, dt$ where $x' = Ax + By$, $x(0) = c$.

4.30. Functional Analysis

Let us now apply some rudimentary functional analysis to establish the fact that

$$J(x) = \int_0^T [(x', x') + 2g(x)] \, dt \tag{4.30.1}$$

attains its minimum for a unique function if $g(x)$ is strictly convex in x and the function $x(t)$ ranges over the function class defined by

(a) $x(0) = c$,   (4.30.2)

(b) $x' \in L^2(0, T)$.

Since $g(x)$ is strictly convex, $g(x)$ has a finite minimum in x. Hence,

$$J(x) \geq a_1 > -\infty \tag{4.30.3}$$

for all x. Thus, lim inf $J(x)$ exists for x in the class of functions described by (4.30.2). Let

$$a_2 = \lim \inf J(x) \tag{4.30.4}$$

and $\{x_n\}$ be a sequence such that

$$a_2 = \lim_{n \to \infty} J(x_n). \tag{4.30.5}$$

Let us now show that $\{x_n'\}$ converges strongly. We have

$$\int_0^T \left(\frac{x_n' - x_m'}{2}, \frac{x_n' - x_m'}{2} \right) dt + J\left(\frac{x_n + x_m}{2} \right)$$

$$= \int_0^T \left[\left(\frac{x_n' - x_m'}{2}, \frac{x_n' - x_m'}{2} \right) + \left(\frac{x_n' + x_m'}{2}, \frac{x_n' + x_m'}{2} \right) \right] dt$$

$$+ 2 \int_0^T g\left(\frac{x_n + x_m}{2} \right) dt. \tag{4.30.6}$$

Hence,

$$\int_0^T \left(\frac{x_n' - x_m'}{2}, \frac{x_n' - x_m'}{2} \right) dt = \frac{1}{2} \int_0^T [(x_n', x_n') + (x_m', x_m')] \, dt$$

$$+ 2 \int_0^T g\left(\frac{x_n + x_m}{2} \right) dt - J\left(\frac{x_n + x_m}{2} \right). \tag{4.30.7}$$

Now

$$g\left(\frac{x_n + x_m}{2}\right) \le \frac{1}{2}[g(x_n) + g(x_m)], \qquad (4.30.8)$$

by virtue of convexity, and from the definition of a_2,

$$J\left(\frac{x_n + x_m}{2}\right) \ge a_2. \qquad (4.30.9)$$

Hence, (4.30.7) yields

$$\int_0^T \left(\frac{x_n' - x_m'}{2}, \frac{x_n' - x_m'}{2}\right) dt$$

$$\le \frac{1}{2}\int_0^T [(x_n', x_n') + (x_m', x_m')] \, dt + \frac{1}{2}\int_0^T [2g(x_n) + 2g(x_m)] \, dt - a_2$$

$$= \frac{1}{2}\left\{\int_0^T [(x_n', x_n') + 2g(x_n)] \, dt - a_2\right\}$$

$$+ \frac{1}{2}\left\{\int_0^T [(x_m', x_m') + 2g(x_m)] \, dt - a_2\right\}$$

$$= \tfrac{1}{2}[(J(x_n) - a_2) + (J(x_m) - a_2)]. \qquad (4.30.10)$$

Hence, as $m, n \to \infty$,

$$\int_0^T \left(\frac{x_n' - x_m'}{2}, \frac{x_n' - x_m'}{2}\right) dt \to 0. \qquad (4.30.11)$$

Thus, the sequence $\{x_n'\}$ converges uniformly to a function $y(t)$ as $n \to \infty$. Since

$$x_n = c + \int_0^t x_n' \, dt_1, \qquad (4.30.12)$$

it follows that x_n also converges strongly to a function $z(t)$. Since

$$z = c + \int_0^t y \, dt_1, \qquad (4.30.13)$$

we see that $y = z'$.

4.31.   Existence and Uniqueness of Solution of Euler Equation

We are now in a position to demonstrate the existence and the uniqueness of the solution of the two-point, boundary-value problem

$$x'' - \operatorname{grad} g = 0,$$
$$x(0) = c, \qquad x'(T) = 0, \qquad (4.31.1)$$

provided that g is strictly convex. The existence of a solution follows from the fact that the minimizing function z determined above must satisfy (4.31.1). We obtain this result using the method of Haar given in Section 4.5. The uniqueness follows from the fact that any solution of (4.31.1) must provide the absolute minimum of $J(x)$, since $J(x)$ is a strictly convex functional.

In the next chapter we shall discuss some aspects of the effective determination of x. This is a formidable problem, and we are nowhere near its solution. We do possess some interesting methods which resolve subclasses of problems.

4.32.   Global Constraints

Let us now suppose that we wish to minimize the functional

$$J(u) = \int_0^T g(u, u') \, dt \qquad (4.32.1)$$

subject to the initial-value condition $u(0) = c$ and a global constraint of the form

$$\int_0^T k(u, u') \, dt = a_1. \qquad (4.32.2)$$

We are tempted to proceed in the following fashion. Introduce a Lagrange parameter λ and consider the new functional

$$J(u, \lambda) = \int_0^T [g(u, u') + \lambda k(u, u')] \, dt. \qquad (4.32.3)$$

The Euler equation

$$(g_{u'} + \lambda k_{u'})' - (g_u + \lambda k_u) = 0,$$
$$u(0) = c, \qquad (g_{u'} + \lambda k_{u'})_{t=T} = 0, \qquad (4.32.4)$$

determines, presumably, a function of t and λ, $u(t, \lambda)$. Let λ be determined to satisfy (4.32.2), i.e., by the equation

$$\int_0^T k(u(t, \lambda), u'(t, \lambda)) \, dt = a_1. \qquad (4.32.5)$$

Then $u(t, \lambda)$, with λ fixed in this fashion, is the desired solution.

As we mentioned in the chapter on discrete control processes, there are serious difficulties in validating this approach. Nonetheless, it is easy to show that the method is valid when it works. By this we mean that if we can find a function in the foregoing fashion, then it does minimize J subject to (4.32.2); see the discussion in Chapter 2 and in Volume I.

EXERCISE

1. Minimize $\int_0^T u'^2 \, dt$ subject to $\int_0^T u^2 \, dt = 1$, $u(0) = c$.

4.33. Necessity of Euler Equation

It is not difficult to show that the equation in (4.32.4) is a necessary condition. We begin with

$$J(u + \varepsilon \bar{u}) = \int_0^T g(u + \varepsilon \bar{u}, u' + \varepsilon \bar{u}') \, dt$$

$$= J(u) + \varepsilon \int_0^T [\bar{u} g_u + u' g_{u'}] \, dt + 0(\varepsilon^2), \qquad (4.33.1)$$

and derive the customary variational equation

$$\int_0^T [\bar{u} g_u + \bar{u}' g_{u'}] \, dt = 0. \qquad (4.33.2)$$

The global constraint in (4.32.2) furnishes a further variational condition

$$\int_0^T [\bar{u} k_u + \bar{u}' k_{u'}] \, dt = 0. \qquad (4.33.3)$$

Integrating by parts in both expressions, we have

(a) $\displaystyle \int_0^T \left[\int_t^T g_u \, dt_1 + g_{u'} \right] \bar{u}' \, dt = 0,$

$\qquad\qquad\qquad\qquad\qquad\qquad\qquad\qquad (4.33.4)$

(b) $\displaystyle \int_0^T \left[\int_t^T k_u \, dt_1 + k_{u'} \right] \bar{u}' \, dt = 0,$

for all $\bar{u}' \in L^2(0, T)$. The relation in (4.33.4a) must hold for all $\bar{u}$ satisfying (4.33.4b).

At this point we invoke a lemma:

LEMMA. *If*

$$\int_0^T gv \, dt = 0 \qquad (4.33.5)$$

for all $v \in L^2 (0, T)$ such that

$$\int_0^T fv \, dt = 0, \qquad (4.33.6)$$

where $f, g \in L^2(0, T)$, then

$$g = \lambda f \qquad (4.33.7)$$

almost everywhere, for some constant λ.

To prove the lemma, consider the quadratic in λ,

$$\int_0^T (g - \lambda f)^2 \, dt. \tag{4.33.8}$$

The minimizing value of λ is furnished by the equation

$$\int_0^T (g - \lambda f)f \, dt = 0. \tag{4.33.9}$$

Regarding $g - \lambda f$ as a function v, we see that we must then have, by hypothesis,

$$\int_0^T (g - \lambda f)g \, dt = 0. \tag{4.33.10}$$

The two relations (4.33.9) and (4.33.10) yield

$$\int_0^T (g - \lambda f)^2 \, dt = 0, \tag{4.33.11}$$

whence the conclusion.

Using this result in (4.33.4), we see that a constant λ exists such that

$$\left[\int_t^T g_u \, dt_1 + g_{u'} \right] = \lambda \left[\int_t^T k_u \, dt_1 + k_{u'} \right]. \tag{4.33.12}$$

This is one origin of the Lagrange parameter.

4.34.   Minimization by Inequalities

The use of the calculus of variations to treat variational problems involving global constraints introduces so many analytic complications that there is considerable motivation to develop alternate techniques. In Volume I we indicated one application of the theory of inequalities. Now, let us present another application.

The determination of an optimum transversal contour leads to the minimization of the drag integral

$$J(u) = \int_0^{2\pi} [u^6/(u^2 + \dot{u}^2) + 2(u^2 + \dot{u}^2)^{1/2}] \, dt, \tag{4.34.1}$$

subject to the constraint

$$\int_0^{2\pi} u^2 \, dt = a_1. \tag{4.34.2}$$

Rather than approach this in a conventional way by using the method outlined above, let us employ a basic inequality due to Young. This reads

$$a^3 + 2b^{3/2} \geq 3ab \tag{4.34.3}$$

for $a, b \geq 0$. Comparing areas in Figure 4.4, the result is immediate. Equality occurs if and only if $b = a^2$.

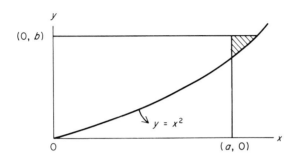

Figure 4.4

If we write $a = f(t)$, $b = g(t)$, and integrate over $[0, 2\pi]$, we obtain the integral inequality

$$\int_0^{2\pi} f^3 \, dt + 2 \int_0^{2\pi} g^{3/2} \, dt \geq 3 \int_0^{2\pi} fg \, dt \tag{4.34.4}$$

for any two nonnegative functions f and g for which the left side exists, with equality only if $g = f^2$.

Write

$$f = u^2/(u^2 + u'^2)^{1/3}, \qquad g = (u^2 + u'^2)^{1/3}, \tag{4.34.5}$$

then $fg = u^2$, and (4.34.4) yields

$$J(u) = \int_0^{2\pi} [u^6/(u^2 + u'^2) + 2(u^2 + u'^2)^{1/2}] \, dt \geq 3 \int_0^{2\pi} u^2 \, dt \tag{4.34.6}$$

with equality only if

$$(u^2 + u'^2)^{1/3} = u^4(u^2 + u'^2)^{-2/3}. \tag{4.34.7}$$

EXERCISE

1. Determine the solution to the original minimization problem starting from (4.34.7).

4.35.   Inverse Problems

Suppose we are given a second-order equation

$$u'' = k(u, u'),$$

or a relation between u and u',

$$u' = p(u, t).$$

When can we consider these to be equivalent to an Euler equation of some functional? In other words, if we observe a specific type of feedback control in a system of biological, economic, or engineering type, when can we assert that there is an appropriate criterion function for which this observed behavior represents an optimal policy?

These are called *inverse problems*. As we shall see in Chapter 6, they can be conveniently treated by means of dynamic programming.

Miscellaneous Exercises

1. Consider the problem of minimizing $J(u) = \int_0^T [u'^2 + q(t)u^2 + 2f(t)u]\, dt$ where $\int_0^T [u'^2 + q(t)u^2]\, dt$ is positive definite and $u(0) = u(T) = 0$. Show that if $f(t) \geq 0$ then $u \leq 0$, where u is the minimizing function.

2. Establish the same result for $J(u) = \int_0^T [u'^2 + u^2 + u^4 + 2f(t)u]\, dt$, where again $u(0) = u(T) = 0$.

3. Consider the problem of minimizing a functional of the form $J(u) = \int_0^T g\left(u, u', \int_0^T h(u, u')\, dt_1\right) dt$. (See

    R. Bellman, "Functional Equations in the Theory of Dynamic Programming—XV: Layered Functionals and Partial Differential Equations," *J. Math. Anal. Appl.*, **28**, 1969, pp. 1–3.
    D. G. B. Edelen, "Problems of Stationanity and Control with a Nonlocal Cost Per Unit Time," *JMAA*, **28**, 1969, pp. 660–673.)

4. Show that the Euler equations associated with $J(u) = \int_0^\infty (u'^2 + 4u^{5/2}t^{-1/2}/5)\, dt$, $u(0) = 1$, is $u'' - u^{3/2}t^{-1/2} = 0$, the Fermi-Thomas equation.

5. Write $f(a, c) = \min_u \int_a^\infty (u'^2 + u^m t^{-n})\, dt$, $u(a) = c$, $u(\infty) = 0$. Then $f_a = (f_c)^2/4 - c^m a^{-n}$.

6. Show that $f(a, c) = a^{(2+m-n)/(2-m)}f(1, c/a^{(2-n)/(2-m)})$, $m \neq 2$, and thus obtain an ordinary differential equation for $f(1, t)$.

7. Consider similarly $J(u) = \int_a^\infty (u'^2 + e^{\lambda t}u^m)\, dt$, $u(a) = c$. See

R. Bellman, "Dynamic Programming and the Variational Solution of the Fermi-Thomas Equation," *J. Phys. Soc. Japan*, **12**, 1967, pp. 1049–1050.

8. Consider the problem of minimizing $\int_{-\infty}^{\infty} (u'^2 + u' * u') \, dt$, $u(0) = 0$, $u(1) = 1$, where $u' * u' = \int_{-\infty}^{\infty} u(s)v(t - s) \, ds$.

(D. A. Sanchez, "On Extremals for Integrals Depending on a Convolution Product," *J. Math. Anal. Appl.*, **11**, 1965, pp. 213–216.)

9. Consider the equation $u'' - f(t)u = 0$, $u(0) = 1$, $u'(0) = 0$. Maximize $u(1)$ subject to the constraint $\int_0^T f^2(t) \, dt = m$. (The calculation of variations affords an opportunity to obtain precise bounds in stability theory. For detailed analysis of this nature, see

N. G. de Bruijn, "A Problem of Optimal Control," *J. Math. Anal. Appl.*, **14**, 1966, pp. 185–190.
A. M. Fink, "Maximum Amplitude of Controlled Oscillations," *J. Math. Anal. Appl.*, **14**, 1966, pp. 253–262.
J. R. Wilkins, Jr., *SIAM Rev.*, **11**, 1969, pp. 86–88.

10. Consider the problem of minimizing $J(u) = \int \left[\sum f_k(t, u, u') \right] dt$, where $f_k(\lambda t, u, \lambda^{-1} u') = \lambda^k f_k(t, u, u')$. Show that $\sum k \int f_k \, dt = 0$ is a necessary condition.

11. If $t^\alpha u'' + \alpha t^{\alpha-1} u' + \gamma t^\beta u^{\gamma-1} = 0$, then $\int t^\beta u^\gamma \, dt = 0$. (See

G. Rosen, "A Necessary Condition for the Existence of Singularity-Free Global Solutions to Nonlinear Ordinary Differential Equations," *Quart. Appl. Math.*, **XXVII**, 1969, pp. 133–134.

12. Consider the problem of minimizing $\int_0^T (u^2 + v^2) \, dt$ subject to $u' = -u + v$, $u(0) = 1$. Suppose that we suboptimize and set $v = au$. Determine the optimal choice of a. What happens as $T \to \infty$?

13. Consider the corresponding multidimensional case.

14. Suppose $x' = Ax + y$, $x(0) = c$, $J(x, y) = \int_0^T [(x, x) + (y, y)] \, dt$, and we want y to depend linearly only on the first k components of x. What is the optimal procedure?

15. What is the connection between minimizing the distance to a given point, or surface, at time T and minimizing the time to arrive at the given point, or surface?

BIBLIOGRAPHY AND COMMENT

4.1. For a detailed discussion of quadratic functionals, see

R. Bellman, *Introduction to the Mathematical Theory of Control Processes*, Vol. I: *Linear Equations and Quadratic Criteria*, Academic Press, New York, 1968.

For the classical theory of the calculus of variations, see

R. Hermann, *Differential Geometry and the Calculus of Variations*, Academic Press, New York, 1968.

M. Morse, *Calculus of Variations in the Large*, American Mathematical Society, 1935.

Some books specifically concerned with the development of this theory for application to control theory are

R. Bellman, I. Glicksberg, and O. Gross, *Some Aspects of the Mathematical Theory of Control Processes*, The RAND Corp. R-313, 1958.

M. Hestenes, *Calculus of Variations and Optimal Control Theory*, Wiley, New York, 1966.

An important work applying the theory of continuous groups to the calculus of variations is

E. Noether, "Invariant Variationsprobleme," *Nachr. Ges.*, Göttingen, 1918, pp. 235–257.

For bibliographical and historical details, see

R. Bellman and R. Kalaba (eds.), *Mathematical Trends in Control Theory*, Dover, New York, 1964.

P. Dorato, *A Selected Bibliography and Summary of Articles on Automatic Control*, Res. Rep. *PIBMRI*-1196-63, Polytechnic Institute, Microwave Research Institute, Jan. 1963.

A. T. Fuller, "Bibliography of Optimum Nonlinear Control of Deterministic and Stochastic-Definite Systems," *J. Electron. Control*," First series, **13**, 1962, pp. 589–611.

For discussions of the origins of many types of control processes in mathematical economics, see

K. J. Arrow, "Applications of Control Theory to Economic Growth," *Lec. Appl. Math.* **12**, pp. 85–119, 1968.

R. Bellman, "Dynamic Programming and Its Application to the Variational Problems in Mathematical Economics, *Proc. Symp. in Calculus of Variations and Appl.* April, 1956, pp. 115–138.

A. R. Dobell, "Some Characteristic Features of Optimal Control Problems in Economic Theory," **AC-14**, 1969, pp. 39–47.

P. Massé, *Les reserves et la regulation de l'avenir dans la vie economique.* Vol. 1: *Avenir determine;* Vol. II: *Avenir aleatoire*, Hermann et Cie, Paris 1946.

We have omitted any discussion of the use of Lyapunov functions to drive a system to a desired equilibrium state, not necessarily in minimum time.

4.12. For references to the extensive theory of convexity, see

E. F. Beckenbach and R. Bellman, *Inequalities*, Springer, Berlin, 1961.

4.20. A simple derivation of Hamilton-Jacobi theory using dynamic programming is given in

S. Dreyfus, *Dynamic Programming and the Calculus of Variations*, Academic Press, New York, 1965.

See the book by Hermann cited above for the conventional derivation.

More general results can be obtained for systems of differential equations not derivable from a variational problem. See

R. Bellman and R. Kalaba, "A Note on Hamilton's Equations and Invariant Imbedding," *Quart. Appl. Math.*, **21**, 1963, pp. 166–168.

R. Bellman and R. Kalaba, "Invariant Imbedding and the Integration of Hamilton's Equations," *Rend. Circ. Mat. Palermo*, **12**, 1963, pp. 1–11.

4.22. We shall have more to say about the return function in Chapter 6. It plays a fundamental role in the dynamic programming treatment of control processes of continuous type.

4.24. For a discussion of minimization of the maximum functional, see

G. Aronsson, "Minimization Problems for the Functional sup $F(x, f(x), f'(x))$," *Ark. Mat.* **6**, 1965, pp. 33–53.

G. Aronsson, "Minimization Problems for the Functional sup $F(x, f(x), f'(x))$—II," *Ark. Mat.* **6**, 1966, pp. 409–431.

C. D. Johnson, "Optimal Control with Chebyshev Minimax Performance Index," *ASME Trans. J. Basic Eng.*, June 1967.

4.30. This is a minor modification of an argument given in the book by Bellman, Glicksberg, and Gross cited in 4.1.

4.33. For a discussion of the geometric significance of the Lagrange multiplier, see the Bellman, Glicksberg, and Gross book. The Lagrange multiplier can also be interpreted as a "price" as we have already indicated; see Chapter 6.

4.34. See

A. Miele (ed.), *Theory of Optimum Aerodynamic Shapes*, Academic Press, New York, 1965,

for further details, and in particular,

R. Bellman, "Young's Inequality and the Problem of the Optimum Transversal Contour," In *Theory of Optimum Aerodynamics Shape* (A. Miele, ed.), Academic Press, New York, 1965, pp. 315–317.

Seven papers worth consulting for additional material not included in the text are

J. H. George and W. G. Sutton, "Application of Liapunov Theory to Boundary-Value Problems," *JMAA*, to appear.

D. Gorman and J. Zaborszky, "Functional Representation of Nonlinear Systems, Interpolation and Lagrange Expansion for Functionals," *ASME Trans.* 1966.

C. Olech, "Existence Theorems for Optimal Problems with Vector-Valued Cost Functions," *Trans. Amer. Math. Soc.*, **136**, 1969, pp. 159–180.

L. I. Rozonoer, "Variational Approach to the Problem of Invariance of Automatic Control Systems—I," *Avtomat. Telemekh.* **24**, 1963, pp. 744–756.

T. L. Vincent and J. D. Mason, "Disconnected Optimal Trajectories," *JOTA*, **3**, 1969, pp. 263–281.

A. Wouk, "An Extension of the Caratheodory-Kalman Variational Method," *J. Optimization Theory Appl.* **3**, 1968, pp. 2–34.

J. Zaborsky and D. Gorman, "Control by Functional Lagrange Expansions," *IFAC*, 1966.

5

COMPUTATIONAL ASPECTS OF THE

CALCULUS OF VARIATIONS

5.1. Introduction

In the preceding chapter it was demonstrated with the aid of various plausible assumptions that the minimization of the functional

$$J(x) = \int_0^T [(x', x') + 2h(x)] \, dt, \tag{5.1.1}$$

subject to the initial condition $x(0) = c$, was equivalent to the solution of the two-point boundary-value problem

$$x'' - \operatorname{grad} h = 0, \tag{5.1.2}$$

$$x(0) = c, \qquad x'(T) = 0.$$

In this chapter we wish to examine the feasibility of this latter task, granted the aid of a digital computer. Despite the power of the contemporary computer, a numerical solution of an equation of this nature is neither simple nor guaranteed. A digital computer is ideally designed to solve initial-value problems. Unfortunately, the value $x'(0)$ is missing, as is $x(T)$. Thus, a straightforward numerical procedure for the calculation of x cannot be applied.

We cannot in general expect any assistance from an explicit analytic solution in terms of the elementary function of analysis because the equation in (5.1.2) is nonlinear in most important circumstances. As analog computers become more powerful, we can expect considerable help from this quarter.

Since, however, this is still a rather unexplored area, we shall restrict our attention here to methods which can be used solely in conjunction with digital computers.

In any case, it must be straightforwardly admitted that the analytic and computational questions posed by multipoint boundary-value problems are difficult and occasionally intransigent. For a number of reasons we cannot expect to discover any single method which will provide a panacea and work uniformly well in all circumstances. Even in treating a particular equation we can anticipate that we will be required to mix techniques in some unanticipated fashion.

In the beginning of this chapter we shall concentrate primarily upon the use of successive approximations, describing some ramifications of both the conventional approach and the theory of quasilinearization. As in Chapter 3, we wish to emphasize the importance of a close examination of all steps in a calculation. Not only is this scrutiny of practical significance, but it possesses the further merit that an appreciation of the serious limitations of all existing methods acts as a constant spur to further research.

In the second half of this chapter we shall briefly sketch some other methods of major importance which can be used and indicate their significance in conjunction with successive approximations. References to more extensive discussions will be found at the end of this chapter.

5.2. Successive Approximations and Storage

Let us consider the simple scalar equation

$$u'' = g(u), \qquad u(0) = c, \qquad u'(T) = 0, \tag{5.2.1}$$

to fix our ideas. We can, of course, remove this from the minute class of equations susceptible to analytic devices by replacing $g(u)$ by $g(u, u')$. Let us, however, retain the simpler version to preserve notational ease. We have previously discussed the convergence of the sequence $\{u_n\}$ determined by the recurrence relation

$$u_n'' = g(u_{n-1}), \qquad u_n(0) = c, \qquad u_n'(T) = 0, \qquad n \geq 1, \tag{5.2.2}$$

with $u_0 = c$. Let us now see what is involved computationally in determining the individual elements of the sequence $\{u_n\}$.

The linearity of the equation for u_n permits us readily to take account of the two-point boundary-value problem in (5.2.2). We have, by a double integration,

$$u_n = c - \int_0^t \left[\int_{t_2}^T g(u_{n-1}) \, dt_1 \right] dt_2. \tag{5.2.3}$$

Thus, $u_n(t)$ for $0 \leq t \leq T$ can be evaluated by numerical quadrature techniques if $u_{n-1}(t)$ is known over $[0, T]$. It is perhaps easier to utilize differential equations directly. Let the function v_{n-1} be defined over $0 \leq t \leq T$ by

$$v'_{n-1} = g(u_{n-1}), \qquad v_{n-1}(T) = 0. \qquad (5.2.4)$$

Then (5.2.3) is equivalent to

$$u'_n = v_{n-1}, \qquad u_n(0) = c. \qquad (5.2.5)$$

We can thus determine $u_n(t)$ by integrating (5.2.4) and (5.2.5) simultaneously, provided that we store the function u_{n-1}. With contemporary computers we anticipate no problem in carrying out this procedure even if an exceptionally fine grid is required.

If, on the other hand, we attempt to apply this procedure to a vector system of dimension one hundred or five hundred with a grid size of one-thousandth or ten-thousandth of the interval length, we can anticipate some difficulties. The requirement for rapid-access storage can well become an onerous burden.

EXERCISES

1. Show that the equation in (5.2.2) is the Euler equation associated with the function

$$J(u, u_n) = \int_0^T (u'^2 + 2h(u_n) + 2(u - u_n)h'(u_n)) \, dt.$$

2. If $h(u)$ is convex, show that

$$J(u) = \int_0^T (u'^2 + 2h(u)) \, dt \geq \int_0^T (u'^2 + 2h(u_n) + 2(u - u_n)h'(u_n)) \, dt.$$

3. Hence, show that

$$\min_u J(u) \geq \min_u \left[\int_0^T (u'^2 + 2h(u_n) + 2(u - u_n)h'(u_n)) \, dt \right],$$

and thus that

$$J(u_n) \geq \min_u J(u) \geq \left[\int_0^T (u_{n+1}'^2 + 2h(u_n) + 2(u_{n+1} - u_n)h'(u_n)) \right] dt.$$

5.3. Circumvention of Storage

Fortunately, the requirement of large amounts of rapid-access storage for the algorithm presented in Section 5.2 can be circumvented in the following fashion. Consider the equation for u_1,

$$u''_1 = g(u_0), \qquad u_1(0) = c, \qquad u'_1(T) = 0. \qquad (5.3.1)$$

Then

$$u_1''(0) = - \int_t^T g(u_0) \, dt_1.$$ (5.3.2)

Hence, the missing initial condition is given by

$$u_1'(0) = - \int_0^T g(u_0) \, dt_1 = b_1.$$ (5.3.3)

This value, b_1, can be calculated as the solution of the initial-value problem

$$v_1' = -g(u_0), \qquad v_1(0) = 0,$$ (5.3.4)

at the point $t = T$. In this case, of course, the solution is immediate, $v_1(T) = -Tg(c)$. Once b_1 has been evaluated, we can replace the two-point boundary-value problem of (5.3.1) by the more tractable initial-value problem

$$u_1'' = g(u_0), \qquad u_1(0) = c, \qquad u_1'(0) = b_1.$$ (5.3.5)

Consider next the determination of u_2 using the equation

$$u_2'' = g(u_1), \qquad u_2(0) = c, \qquad u_2'(T) = 0.$$ (5.3.6)

As before, the missing initial condition $u_2'(0)$ can be determined as the solution of

$$v_2' = -g(u_1), \qquad v_2(0) = 0,$$ (5.3.7)

evaluated at $t = T$. To calculate v_2 in this fashion, however, we require the function u_1, and this appears to demand storage of the function u_1 over $[0, T]$. To avoid this, we adjoin the equation in (5.3.5) to the equation in (5.3.7) and consider the simultaneous system

$$u_1'' = g(u_0), \qquad u_1(0) = c, \qquad u'(0) = b_1,$$
$$v_2' = -g(u_1), \qquad v_2(0) = 0,$$ (5.3.8)

where $u_0 = c$. In place of storing the functional values of u_1 for $0 \le t \le T$, we generate the desired values as we go along.

Numerical integration of this system yields the missing initial condition

$$v_2(T) = u_2'(0) = b_2.$$ (5.3.9)

The equation for the determination of u_2 in (5.3.6) is then replaced by the initial-value equation

$$u_2'' = g(u_1), \qquad u_2(0) = c, \qquad u_2'(0) = b_2.$$ (5.3.10)

When we turn to the determination of $u_3'(0)$ by means of the equation

$$u_3'' = g(u_2), \qquad u_3(0) = c, \qquad u_3'(T) = 0,$$ (5.3.11)

we must adjoin two equations, (5.3.5) and (5.3.10), to avoid storage of either u_1 or u_2. Continuing in this fashion, we can step-by-step obtain $u_n(t)$ without ever having to store $u_1, u_2, \ldots, u_{n-1}$.

EXERCISE

1. How many initial-value problems must be solved to obtain u_N in the foregoing fashion?

5.4.   Discussion

Bypassing the storage of the previous approximation in the calculation of the next approximation permits us to obtain a very high degree of accuracy solely at a cost in time by taking a sufficiently small step size in the numerical integration of the differential equations. On the other hand, new difficulties can arise in the application of this method to systems of high dimension due to the large number of differential equations of initial-value type appearing at the n-th stage.

There is thus considerable motivation for reducing the number of steps required in the method of successive approximations. We shall consider this question below.

EXERCISES

1.   Consider the differential-difference equation

$$u'(t) = g(u(t), u(t-1)), \quad t \geq 1,$$

$$u(t) = h(t), \quad 0 \leq t \leq 1.$$

Discuss techniques of numerical solution of this equation involving the storage of functions over intervals of length one.

2.   Show that the preceding equation is equivalent to an infinite system of ordinary differential equations

$$v_n'(t) = g(v_n(t), v_{n-1}(t)), \quad n \geq 1,$$

where t is now restricted to the interval $[0, 1]$ and $v_0 = h(t)$. (*Hint*: Consider the sequence $\{v_n(t)\}$ determined by $v_n(t) = u(t+n), 0 \leq t \leq 1$.)

3.   Show that the equation in Exercise 1 can be solved numerically without the storage of functions in the case where $h(t)$ is a function not requiring storage, e.g., a polynomial or a solution of a differential equation. (*Hint*: First solve the equation $v_1' = g(v_1, v_0), v_1(0) = h(1), v_0 = h(t)$. This determines $v_1(1)$. Then solve $v_1' = g(v_1, v_0)$, $v_1(0) = h(1)$, $v_2' = g(v_2, v_1)$, $v_2(0) = v_1(1)$, $v_0 = h(t)$, etc.)

4. Show that we can obtain the solution of $u'' = (a + b \cos t)u = 0$ computationally without storing the values of $\cos t$.

5.5. Functions and Algorithms

Once again we have been emphasizing in the foregoing sections and associated exercises the fact that a function may be determined by any of a number of different algorithms. Accordingly, only those algorithms for the determination of functional values which are most convenient in a particular situation should be employed in the course of the numerical calculation.

In the foregoing case, we use an algorithm which exchanges time for rapid-access storage capacity. Observe that in so doing we discard an enormous amount of data at each stage, retaining only the small amount of data which is absolutely required to generate the same total quantity of original data, and more, at the next stage of the computational process. This essential residue of data can correctly be described as "information." The function is then described by a set of data plus an algorithm for generating additional data as desired. Observe how the very concept of information is dependent upon the way in which the crucial data is utilized.

As the analog computer, and particularly as the hybrid computer—part digital and part analog—becomes more powerful, it is quite probable that we will reverse the direction of a great deal of current research. Rather than seeking to convert two-point, and multipoint, boundary-value problems into initial-value problems, we will seek to utilize the almost instantaneous solution of large boundary-value problems by using analog computers to resolve many types of questions concerning the time history of a complex process. This is intimately connected with the difficult area of "on-line control" which we will not investigate in this volume.

Furthermore, in order to ensure accuracy and, more importantly, success in a computation, we will employ several different types of methods in parallel, or in series-parallel, order. Some of these methods will be purely digital, some purely analog, and some hybrid. Many important scheduling problems arise in this area, some of which can be treated by means of dynamic programming.

Above all, we wish to promulgate the twin ideas of flexibility and versatility in methodology, eschewing the easily acquired bad habit of naive and rigid adherence to a particular method. No single theory, irrespective of intrinsic elegance and regardless of a record of successes prepared by its champions, is sufficiently powerful to resolve more than a small fraction of outstanding questions of significance in control theory. Complex questions in all areas require for their solution a broad cultural basis. Mathematics is no exception.

5.6. Quasilinearization

In our preliminary discussion of some computational aspects of the method of successive approximations, we indicated that the technique we employed to avoid the storage of functions can lead to the task of solving uncomfortably large systems of simultaneous ordinary differential equations because of the relatively slow convergence of the sequence of functions generated by the straightforward approach proposed.

Let us see if we can construct an alternate approximation scheme which converges much more rapidly than that used in the previous sections and thus avoids as rapid a buildup of the number of equations that must be solved simultaneously. In place of the relation

$$u_n'' = g(u_{n-1}), \tag{5.6.1}$$

which approximates to the correct expression $g(u_n)$ by using the first term in the Taylor expansion around u_{n-1}, let us examine the relation

$$u_n'' = g(u_{n-1}) + (u_n - u_{n-1})g'(u_{n-1}) + \cdots + \frac{(u_n - u_{n-1})^k}{k!}\, g^{(k)}(u_{n-1}), \tag{5.6.2}$$

obtained using the first $k + 1$ terms. It is intuitively clear that we will obtain a better approximation using a recurrence relation of this nature. A rigorous proof requires some ideas of stability theory of the type expounded in Chapter 8. If $k > 1$, however, we possess no routine way of determining u_n since the equation for u_n is nonlinear and subject to the original two point condition. We are hardly better off than when we started.

Let us then compromise. To obtain higher-order approximation without sacrificing linearity, we take $k = 1$. This leads to the relation

$$u_n'' = g(u_{n-1}) + (u_n - u_{n-1})g'(u_{n-1}),$$
$$u_n(0) = c, \qquad u_n'(T) = 0. \tag{5.6.3}$$

Another way of looking at this is that we are using a quadratic approximation in the criterion functional in place of the linear approximation yielding (5.6.1); see the exercises at the end of Section 5.2. If

$$J(u) = \int_0^T [u'^2 + 2h(u)]\, dt, \tag{5.6.4}$$

and if we use the approximation

$$h(u) \cong h(u_{n-1}) + (u - u_{n-1})h'(u_{n-1}) + \frac{(u - u_{n-1})^2}{2}\, h''(u_{n-1}) \tag{5.6.5}$$

in the integrand, we obtain (5.6.3) as the corresponding Euler equation.

This tells us the equation in (5.6.3) possesses a unique solution for u_n whenever the quadratic functional possesses an absolute minimum. One simple condition for this is that $h'' > 0$, convexity; another is that $T \ll 1$. We shall need the latter to establish convergences in general, but we shall return to the former condition.

EXERCISES

1. What happens if we attempt to use the approximate functional $J(u) = \int_0^T [u'^2 + 2h(u_{n-1})] \, dt$ as a way of generating successive approximations?

5.7. Convergence

There are several ways in which we can study the convergence of the sequence $\{u_n\}$. Let us pursue a simple route since we are primarily concerned here with the basic idea of accelerated convergence rather than with establishing any particularly sparkling estimates of the interval of convergence.

Write the solution of (5.6.3) as the solution of the integral equation

$$u_n = c + \int_0^T k(t, t_1)[g(u_{n-1}) + (u_n - u_{n-1})g'(u_{n-1})] \, dt_1. \qquad (5.7.1)$$

We have already employed this Green's function in Chapter 4, together with the simple estimate

$$|k(t, t_1)| \le T, \qquad 0 \le t, \quad t_1 \le T. \qquad (5.7.2)$$

To simplify the subsequent analysis, let us write (5.7.1) in the form

$$u_n = c + S[g(u_{n-1}) + (u_n - u_{n-1})g'(u_{n-1})], \qquad (5.7.3)$$

where S is the linear operation defined in (5.7.1). Furthermore, let us write

$$\|u\| = \max_{0 \le t \le T} |u|. \qquad (5.7.4)$$

It follows from (5.7.2) that if

$$v = S(u), \qquad (5.7.5)$$

then

$$\|v\| \le T^2 \|u\|. \qquad (5.7.6)$$

With these preliminaries disposed of, let us begin the proof of convergence of u_n following the usual path. We start by establishing inductively that $\|u_n\| \le 2|c|$ for $T \ll 1$. We have, assuming the result for $n - 1$,

$$\|u_n\| \le |c| + T^2 \|g(u_{n-1}) - u_{n-1}g'(u_{n-1})\| + T^2 \|u_n g'(u_{n-1})\|. \qquad (5.7.7)$$

Hence,

$$\|u_n\| \le |c| + T^2 k_1 + T^2 k_2 \|u_n\|, \qquad (5.7.8)$$

where

$$
\begin{aligned}
k_1 &= \max_{|u| \le 2|c|} |g(u) - u g'(u)|, \\
k_2 &= \max_{|u| \le 2|c|} |g'(u)|.
\end{aligned}
\qquad (5.7.9)
$$

Thus,

$$\|u_n\| \le \frac{|c| + T^2 k_1}{1 - T^2 k_2} \le |c| \qquad (5.7.10)$$

for T small enough.

To establish convergence, we write

$$
\begin{aligned}
u_{n+1} - u_n &= S[g(u_n) + (u_{n+1} - u_n) g'(u_n)] - S[g(u_{n-1}) + (u_n - u_{n-1}) g'(u_{n-1})] \\
&= S[(u_{n+1} - u_n) g'(u_n)] + S[g(u_n) - g(u_{n-1}) - (u_n - u_{n-1}) g'(u_{n-1})] \\
&= S[(u_{n+1} - u_n) g'(u_n)] + S[(u_n - u_{n-1})^2 g''(\theta)/2],
\end{aligned}
$$
$$(5.7.11)$$

where θ is in $[u_n, u_{n-1}]$ and $k_3 = (\max_{|u| \le 2|c|} |g''(u)|)/2$. Thus, we have

$$
\begin{aligned}
\|u_{n+1} - u_n\| &\le k_2 T^2 \|u_{n+1} - u_n\| + k_3 T^2 \|u_n - u_{n-1}\|^2, \\
\|u_{n+1} - u_n\| &\le \frac{k_3 T^2}{(1 - k_2 T^2)} \|u_n - u_{n-1}\|^2, \quad n \ge 1.
\end{aligned}
\qquad (3.7.12)
$$

Furthermore, the same argument shows that

$$\|u_1 - c\| \le T^2 |g(c)|/(1 - k_2 T^2) = b |g(c)|. \qquad (5.7.13)$$

where

$$b = \max[T^2/(1 - k_2 T^2, k_3 T^2/(1 - k_2 T^2]$$

For $T \ll 1$, we have $b < 1$, and $b |g(c)| < 1$. Thus,

$$
\begin{aligned}
\|u_2 - u_1\| &\le b \|u_1 - c\|^2 \le b(b^2 |g|^2) = b^3 |g|^2, \\
\|u_3 - u_2\| &\le b \|u_2 - u_1\|^2 \le b(b^3 |g|^2)^2 = b^7 |g|^4 \\
&\;\;\vdots \\
\|u_{n+1} - u_n\| &\le b^{2^{n+1}-1} |g|^{2^n}.
\end{aligned}
\qquad (5.7.14)
$$

We see then that the series $\sum_n \|u_{n+1} - u_n\|$ is majorized not by a series of the usual form $\sum_n r^n$ but by a series of the form $\sum_n r^{2^n}$, with $r < 1$, provided that $T \ll 1$. This is called *quadratic convergence*. The convergence is very much more rapid.

EXERCISES

1.   If the sequence defined by (5.7.1) converges uniformly in $[0, T]$, it converges to a solution of the original differential equation.
2.   If the sequence converges uniformly, must it converge quadratically?
3.   Can the condition of uniform convergence be replaced by bounded convergence?

5.8.   Discussion

We have established convergence under the condition that $T \ll 1$. If $h(u)$ is convex, which is to say if $g(u)$ is monotone, the sequence $\{u_n\}$ is uniquely defined for all n for any $T \geq 0$. Consequently, in this case we can always carry out the algorithm for a number of stages in the hopes that it will work. Needless to say this is not a very satisfactory state of affairs, although in practice we are often reduced to this "program and a prayer" approach.

5.9.   Judicious Choice of Initial Approximation

The question arises as to whether we can improve the estimate for the interval of convergence by an inspired choice of an initial approximation. With this objective in mind, let us examine the simple method of successive approximations investigated in Chapter 4 and Section 4.2.,

$$u_{n+1} = c + \int_0^T k(t, t_1)g(u_n) \, dt_1. \tag{5.9.1}$$

We have

$$\|u - u_{n+1}\| = \left\| \int_0^T k(t, t_1)(g(u) - g(u_n)) \, dt_1 \right\|$$

$$\leq k_7 \|g(u) - g(u_n)\| \leq k_8 \|u - u_n\|. \tag{5.9.2}$$

Regardless then of how close u_n is to u over $[0, T]$, there is no guarantee that u_{n+1} is closer to u if T is sizeable. We know from the explicit representation of $k(t, t_1)$ that k_7 increases rapidly with T.

On the other hand, if we employ quasilinearization, we have a result of the form of (5.7.14). Regardless of the size of T, we see that if $\|u - u_0\|$ is sufficiently small, we can ensure that

$$\|u - u_1\| \leq k_9 \|u - u_0\|, \tag{5.9.3}$$

where $k_9 < 1$, and, generally, that $\|u - u_{n+1}\| \leq k_9 \|u - u_n\|$.

Hence, if in some fashion we can obtain an initial approximation which is close to u'' we can guarantee convergence. It is this consideration which motivates the selection of topics in the second half of this chapter. The foregoing discussion illustrates a point we made at the beginning of this chapter, namely that a combination of methods may often be required to obtain a solution to a two-point boundary-value problem. One method provides us with a suitable initial approximation; the second method is a vernier method which supplies the desired accuracy. Numerical solution itself becomes a multistage process in which the order of the individual steps is vital and in which successive steps depend upon the results of the initial steps.

EXERCISE

1.   Do we retain quadratic convergence if $\|u - u_0\| \ll 1$?

5.10.   Circumvention of Storage

As in the case of the lower-order method of successive approximations, we can avoid the storage of a function over $[0, T]$ by increasing the number of differential equations integrated at each step. The details are a bit more irksome, but the basic idea is the same as that sketched in Section 5.3.

5.11.   Multidimensional Case

The method of quasilinearization applied to the vector equation

$$x'' = g(x), \tag{5.11.1}$$

$$x(0) = c, \qquad x'(T) = 0,$$

yields the sequence of equations

$$x''_{n+1} = g(x_n) + J(x_n)(x_{n+1} - x_n),$$

$$x_{n+1}(0) = c, \qquad x'_{n+1}(T) = 0, \tag{5.11.2}$$

$n \geq 0$, with $x_0 = c$, where J is the Jacobian matrix of g evaluated at x_n.

As we have already mentioned, the sequence $\{x_n\}$ is well-defined if $T \ll 1$ or if $h(x)$, the function appearing in the criterion functional, is convex for all x. This condition ensures that J is positive definite for all x. Although there is little difficulty in obtaining analogous analytic results concerning the convergence of x_n, there are serious questions as to the computational feasibility of this algorithm. Let us examine the details of the calculation with a critical eye.

5.12.   Two-point Boundary-value Problems

Referring to (5.11.2), we see that to calculate x_{n+1} we must solve a linear equation of the form

$$y'' - A(t)y = f(t),$$
$$y(0) = c, \qquad y'(T) = 0; \tag{5.12.1}$$

here $A(t) = J(x_n), f(t) = g(x_n) - J(x_n)x_n$.

Let z satisfy the equation

$$z'' - A(t)z = f(t),$$
$$z(0) = z'(0) = 0, \tag{5.12.2}$$

a determination by means of initial values, and set

$$y = z + w. \tag{5.12.3}$$

Then w satisfies the homogeneous equation

$$w'' - A(t)w = 0,$$
$$w(0) = c, \qquad w'(T) = -z'(T) = d. \tag{5.12.4}$$

Let us see what is involved in determining w. Take X_1 and X_2 to be the principal solutions of the matrix equation

$$X'' - A(t)X = 0, \tag{5.12.5}$$

which is to say that X_1 and X_2 are respectively determined by the initial conditions

$$X_1(0) = I, \qquad X_1'(0) = 0,$$
$$X_2(0) = 0, \qquad X_2'(0) = I. \tag{5.12.6}$$

The general solution of (5.12.4) subject to $w(0) = c$ is then given by

$$w = X_1(t)c + X_2(t)b, \tag{5.12.7}$$

where the constant b is to be determined by the remaining condition

$$w'(T) = d. \tag{5.12.8}$$

It follows that b satisfies the linear algebraic equation

$$d = X_1'(T)c + X_2'(T)b, \tag{5.12.9}$$

whence

$$b = X_2'(T)^{-1}[-X_1'(T)c + d]. \tag{5.12.10}$$

We assume that we have imposed conditions that ensure the existence of $X_2'(T)^{-1}$. In general, we require that $T \ll 1$. The condition, however, that $A(t)$ is positive definite for $t \geq 0$ guarantees the existence of $X_2'(T)^{-1}$ for all $T \geq 0$. Once b has been obtained in this fashion, the function $w(t)$ for $t > 0$ is determined by means of an analytic, or numerical, integration of the equation

$$w'' - A(t)w = 0, \qquad w(0) = c, \qquad w'(0) = b; \qquad (5.12.11)$$

analytic if $A(t)$ is constant, numerical otherwise. This is an initial-value problem and therefore presumably well-suited to the abilities of the digital computer. As we shall see, the word "presumably" is well chosen. The probability of success is high, but success is not guaranteed.

5.13.   Analysis of Computational Procedure

Let us now carefully examine the various steps in the procedure just described. There are four major stages:

(a)   the determination of the vector $X_1'(T)c$,
(b)   the determination of the matrix $X_2'(T)$,
(c)   the inversion of the matrix $X_2'(T)$,
(d)   the numerical integration of the initial-value problem in (5.12.11).

As might be expected, whatever difficulties exist magnify as the dimension of x increases. However, even in the case of low dimension, surprising complexities can arise. These obstacles can destroy the efficacy of quasilinearization. We will discuss these matters in individual sections.

5.14.   Instability

The difficulties associated with (a), (b) and (d) in Section 5.13 are all of the same type—the integration of an initial-value problem. This is a bit shocking since we have been laboring mightily to reduce the solution to an operation of this nature. To illustrate, however, the nature of the new obstacles in the path of a numerical solution that arise in this fashion, it is sufficient to take the case where $A(t)$ is constant and positive definite and consider the calculation of $X_2'(T)$.

Let B denote the positive definite square root of A. Then the general solution of

$$x'' - Ax = 0 \qquad (5.14.1)$$

has the form

$$x = e^{Bt}a + e^{-Bt}b, \qquad (5.14.2)$$

where a and b are vectors determined by the initial conditions. In the case of interest in control theory, the initial conditions are such that as T and t increase, we have

$$x \sim e^{-Bt}b^{(1)}. \tag{5.14.3}$$

Let us now see what happens when an equation such as (5.14.1), subject to initial conditions, is integrated numerically. The differential equation is replaced by an approximating difference equation which we can consider to be equivalent to the solution of a new differential equation of the form

$$y'' - Ay = r(t), \qquad y(0) = c^{(1)}, \qquad y'(0) = c^{(2)}. \tag{5.14.4}$$

The function $r(t)$ is a forcing term introduced to represent the result of two different kinds of error: the error inherent in the use of an approximating difference equation and errors due to round-off.

The solution of (5.14.4) will contain the original solution plus additional terms due to the presence of $r(t)$. These terms will contain components from e^{-Bt} and from e^{Bt}, as well. The contribution from e^{-Bt} will have no significant effect on the principal term $e^{-Bt}b^{(1)}$, but the contribution from e^{Bt}, a term of the form $e^{Bt}c^{(3)}$ with $\|c^{(3)}\| \ll 1$, can easily overwhelm the desired solution if $t \gg 1$.

This is an *instability* effect. There are fortunately a number of ways of overcoming or circumventing this phenomenon in many cases, but it cannot be neglected. References to some of these procedures will be found at the end of this chapter; see also Section 5.18.

5.15. Dimensionality

The determination of $X_2'(T)$ requires the numerical integration of the matrix equation

$$X'' - AX = 0, \qquad X(0) = 0, \qquad X'(0) = I, \tag{5.15.1}$$

a system of N^2 simultaneous differential equations, where N is the dimension of the vector x. If $N \gg 1$, we cannot handle the entire system at one time. We can, however, take advantage of the linearity of (5.15.1) and decompose this set of equations, determining $X_2'(T)$ column by column. As usual, we trade time for rapid-access storage capacity. We shall return to this question of decomposition of large systems in Chapter 13.

5.16. Matrix Inversion

The issue of matrix inversion is always a touchy one, particularly when the matrix is of high degree. Nonetheless, the process must always be examined carefully even if N is small. Many examples show the errors that can arise in this way.

There are a number of methods available which are effective provided that the matrix under consideration is not ill-conditioned. Unfortunately, if $T \gg 1$, the matrix $X'_2(T)$ is ill-conditioned. When we say that A is "ill-conditioned," we mean that the solution of $Ax = b$ is extraordinarily sensitive to small errors in b. Another way of looking at this is to say that A^{-1} contains many terms of large magnitude and opposing sign. Small changes of appropriate sign can thus create large effects.

The reason for the ill-conditioning of $X'_2(T)$ is the following. The expression for $X_2(t)$ is

$$X_2(t) = (B^{-1}/2)(e^{Bt} - e^{-Bt}), \qquad (5.16.1)$$

where B is as before. Let $\mu_1 > \mu_2 > \cdots > \mu_N > 0$ be the N characteristic roots of t, taken distinct for simplicity. Then the canonical representation of the symmetric matrix B shows that as t increases, we have

$$X_2(t) \sim e^{\mu_1 t} A_1, \qquad (5.16.2)$$

where A_1 is a highly singular matrix. Thus, $X'_2(T)$ can be seriously ill-conditioned for $T \gg 1$.

5.17. Tychonov Regularization

The problem of solving an ill-conditioned system of linear algebraic equations

$$Ax = b \qquad (5.17.1)$$

is one that arises in numerous ways in every field that presumes to employ quantitative methods. Although ideally we should be able to reformulate problems so that we are not faced with this onerous task where success is so unpredictable, in practice we often are not adroit enough to find an alternative. Let us then digress and briefly describe some important ideas of Tychonov for treating this problem. Interestingly enough, they fall within the province of control theory. Once again we see that the process of *numerical* solution in a situation where the problem has presumably been completely answered analytically can raise new and interesting types of purely *analytic* questions.

As we have previously indicated, any process of numerical solution may itself be interpreted as a control process where we are trying to minimize the terminal error.

It was pointed out by Cauchy that the solution of (5.17.1) is equivalent to the solution of the problem of minimizing the function

$$\|Ax - b\| \tag{5.17.2}$$

where $\|\cdots\|$ denotes some convenient norm. If we employ as a simple measure of deviation the expression

$$(Ax - b, Ax - b), \tag{5.17.3}$$

we readily obtain the variational equation

$$A^T Ax = A^T b, \tag{5.17.4}$$

where A^T denotes the transpose of A. Unfortunately, this equation may be more ill-conditioned than the original, the equation in (5.17.1).

We need not, of course, use a quadratic norm in the foregoing procedure since we now have available the powerful computational techniques of linear programming and nonlinear programming. However, the basic task is not merely that of finding a vector y for which

$$\|Ay - b\| \le \varepsilon. \tag{5.17.5}$$

This may not be a difficult procedure. As indicated above, the ill-conditioning of A produces an instability in the solution of $Ax = b$. Consequently, even if (5.17.5) holds, there is no guarantee that

$$\|x - y\| \le \varepsilon_1 \tag{5.17.6}$$

where x is the solution of (5.17.1), and ε_1 is of the order of magnitude of ε, or within some other acceptable bound. The vectors for which (5.17.5) holds thus can be considered to belong to an equivalence class. The basic task is that of determining the elements in this class which are sufficiently close to x.

The idea of Tychonov is to introduce a function $\phi(w)$ with the property that the minimization of the new criterion function

$$\|Aw - b\| + \phi(w) \tag{5.17.7}$$

yields a useful approximation to the solution of (5.17.1). We can consider $\phi(w)$ to be a "penalty function" which keeps any vector y satisfying the relation $\|Ay - b\| + \phi(y) \le \varepsilon$ close to x.

In general, it is not too difficult to find a suitable function of this nature since we almost always know much more about the vector x than what is contained in the statement that it is a solution of (5.17.1). The mathematical, or scientific, environment of the underlying problem furnishes this additional constraining information. A constantly challenging question, however, is that of converting this supplementary knowledge into an analytic tool.

Let us give examples of this. In a number of cases, an alternative approach will furnish an approximation to the solution of (5.17.1)—a vector x_0. We may then choose

$$\phi(x) = k \, \|x - x_0\|^2, \tag{5.17.8}$$

where k is a positive parameter whose choice is rather delicate, as we discuss below in the exercises. In other cases, we may use known structural features of the solution to construct $\phi(x)$. This is a "self-consistent" approach.

It is perhaps not necessary to emphasize the fact that we cannot expect any easy or routine treatment of the problem of solving ill-conditioned systems of linear algebraic equations. The solution process is much more of a stochastic control process than a deterministic type.

EXERCISES

1. Show that the minimization of $(Ax - b, Ax - b) + k(x - x_0, x - x_0)$, $k > 0$, leads to the equation $(A^T A + kI)x = (A^T b + kx_0)$. As $k \to 0$, show that the solution of this equation approaches that of $Ax = b$, assuming as we do that A^{-1} exists.
2. Let x_0 be chosen arbitrarily and let x_n, $n \geq 1$, be determined recurrently by means of the equation

$$(A^T A + kI)x_n = (A^T b + kx_{n-1}), \qquad n \geq 1.$$

   Show that $\{x_n\}$ converges to the solution of $Ax = b$.
3. How is the rate of convergence of $\{x_n\}$ determined by the value of k? Discuss advantages and disadvantages of choosing k large or small.
4. Discuss the possibility of using an extrapolation technique in k to determine the solution of $Ax = b$ in terms of the solution of the foregoing minimization problem in Exercise 1.

5.18.   Quadrature Techniques

The solution of the initial-value problem

$$y' = By, \qquad y(0) = c, \tag{5.18.1}$$

can possess the instability features described in the foregoing sections. It also possesses an irritating inefficiency. In order to calculate $y(T)$ in the usual fashion, we must laboriously calculate $y(\Delta)$, $y(2\Delta)$, ..., and so on until we reach $N\Delta = T$. Are there ways of speeding up the calculation by evaluating a smaller number of intermediary values? Let us describe two possible approaches to answering this question.

If B is constant, we can use the Laplace transform to obtain

$$L(y) = (sI - B)^{-1}c, \qquad (5.18.2)$$

where $L(y) = \int_0^\infty e^{-st} y \, dt$. The integral exists for $Re(s) \gg 1$. Using quadrature techniques of the type described in Chapter 12, to invert the Laplace transform numerically, we can use (5.18.2) to obtain values of $y(t)$ at a set of selected values $t_1, t_2, \ldots, t_M$.

If B is variable, we can again use "quadrature" techniques to replace $y'(t)$ by a linear combination of values of $y(t)$ at a set of t-values. Thus, we write

$$y'(t_i) \cong \sum_{j=1}^M a_{ij} y(t_j), \qquad i = 1, 2, \ldots, M. \qquad (5.18.3)$$

There are a variety of methods for determining the a_{ij}.

The differential equation in (5.18.1) is replaced by the system of linear algebraic equations

$$\sum_{j=1}^M a_{ij} y(t_j) = B(t_i)y(t_i), \qquad i = 1, 2, \ldots, M.$$

There are a number of ways of taking account of the initial condition.

EXERCISES

1. Discuss the efficacy of "doubling techniques," i.e., writing $e^{Bt} = (e^{Bt/2})^2$.
2. In calculating $A^N c$, as the solution of $x(n + 1) = Ax(n)$, $x(0) = c$, at $n = N$, is it preferable to calculate Ac, $A(Ac)$, etc., or first A^N and then $A^N c$?
3. Suppose $N = 2^k$ and we calculate A^N by doubling techniques, A^2, A^4, A^8, etc. Is it ever profitable to calculate $A^N c$ by calculating first A^N and then $A^N c$?
4. If N is not a power of 2, how should one calculate A^N?

5.19. "The Proof is in the Program"

We see from the foregoing sections that there is a significant difference between an analytically rigorous treatment and one which is computationally feasible. In general, our lack of knowledge concerning large-scale calculations is such that no algorithm can be considered acceptable until it has been tried in a large variety of circumstances. Even then we have no guarantee that it will work in a novel situation, such as an equation of different structure or with significantly different parameters.

It is for this reason, if no other, that we keep constantly searching for new and improved methods. Only with an overlapping network of theories, devices, strategems and tricks* can we hope to approach control processes of significance.

5.20.   Numerical Solution of Nonlinear Euler Equations

Let us take the Euler equation in the form

$$x'' = g(x), \qquad x(0) = c, \qquad x'(T) = 0, \tag{5.20.1}$$

as before, and turn to a discussion of some entirely different approaches to numerical solution. If g is nonlinear, even the question of existence and uniqueness of solution is a difficult one with no unified treatment. Instead, we have a number of local theories strongly dependent in form on the origin of the equation and the idiosyncrasies of $g(x)$. The area of computational solution is even more chaotic.

Fortunately, with the advent of the digital computer we are in a position to try a number of different approaches one after the other, or even simultaneously, in the hopes that one at least will succeed. As constantly mentioned above, combinations of methods can be expected to be more powerful than individual techniques.

The development of improved means of communication between the mathematician and the computer involving the observation and display of intermediate results has enormously altered the computational scene. The mathematician is now in a favorable position to use adaptive methods of solution. That large-scale calculation, however, will remain more art than science cannot be gainsaid. But art can always be enhanced by knowledge.

5.21.   Interpolation and Search

Let us begin with a rather simple and direct approach which is feasible because of the availability of the digital computer. It is sufficient to discuss the scalar equation

$$u'' = g(u), \qquad u(0) = c_1, \qquad u'(T) = 0 \tag{5.21.1}$$

in order to illustrate the basic ideas.

In place of (5.21.1), consider the initial-value problem

$$u'' = g(u), \qquad u(0) = c_1, \qquad u'(0) = c. \tag{5.21.2}$$

---

* Recall that a method is a trick which works at least twice and that a theory contains at least two methods.

This defines a function of c,

$$u'(T) = h(c). \qquad (5.21.3)$$

The original problem is therefore equivalent to the determination of the value, or values, of c which make $h(c)$ zero.

This approach can be seriously considered as long as it is relatively easy to calculate $h(c)$ for a particular value of c. We can then apply methods of interpolation (polynomial and otherwise), gradient methods, and various sophisticated search techniques to solve the equation $h(c) = 0$.

One difficulty present in the application of all of these methods is the fact that the solution of the Euler equation is basically unstable with respect to determination by initial value techniques, as we pointed out in Section 5.14. Hence, when we try to determine $h(c)$ by means of numerical integration, we can readily encounter serious questions of accuracy. Nonetheless, the method is always worth trying, particularly when the dimension of the system is low.

A constant spectre hovering over the calculation is the nonuniqueness of solution of (5.21.1). This can easily result in the determination of a relative minimum of J rather than an absolute minimum. We always face this possibility when basing our calculations on the Euler equation unless we have some a priori guarantee of uniqueness. Observe then the constant interplay between theory and application. Theory without application is like the smile of the Cheshire cat: application without theory is Blind Man's Buff.

EXERCISES

1. Show that u_c satisfies the linear equation $u_c'' = g_u u_c$, $u_c(0) = 0$, $u_c'(0) = 1$.
2. Under what conditions is $u_c > 0$?
3. Obtain the corresponding equation for u_{cc} and consider the question of deciding when $u_{cc} > 0$.
4. How would one apply the Newton-Raphson method to the solution of the equation $h(c) = 0$?

5.22. Extrapolation

As we know, if $T \ll 1$, it is relatively simple to obtain a numerical solution of the two-point boundary-value problem in (5.19.1) in any of a number of different ways. Can we use this solution to determine the solution for larger values of T in some step-by-step process? The basic idea is to use the solution over $[0, T_0]$ as an approximation to the solution over $[0, T_0 + \Delta]$. If Δ is small, this solution over $[0, T_0]$ will be an excellent first approximation to the solution over $[0, T_0 + \Delta]$, which means that various procedures will succeed and they will succeed quickly. Clearly, the choice of Δ is crucial. If it is too small, the entire process will be too time-consuming; if it is too large, the process may break down due to inaccuracy of the initial approximations. In general, we want the algorithm to be an adaptive one, with Δ dependent not only on T but on various properties of the solution over $[0, T_0]$. We shall not, however, consider this feature here.

There are two immediate approaches. Let

$$c_2(T) = u'(0), \tag{5.22.1}$$

in (5.21.1). Then we can determine a first approximation to u over $[0, T_0 + \Delta]$ by means of the solution of

$$u'' = g(u), \qquad u(0) = c, \qquad u'(0) = c_2(T_0), \tag{5.22.2}$$

where now $0 \le t \le T_0 + \Delta$.

Alternatively, we can begin with the solution of (5.22.2) over $[0, T_0]$ and modify it so that it satisfies the boundary condition at $t = T_0 + \Delta$. For example, we could use the "stretched-out" function

$$u\left(\frac{t(T_0 + \Delta)}{T_0}\right), \qquad 0 \le t \le T_0, \tag{5.22.3}$$

as an initial approximation over $[0, T_0 + \Delta]$.

The missing initial condition $u'(0)$ is a function of c and T, say $f(c, T)$. The study of its dependence upon c and T requires the theory of invariant imbedding for general two-point boundary-value problems. For Euler equations the theory of dynamic programming can be employed; see Chapter 6 for discussion and references.

5.23.   Bubnov-Galerkin Method

Let us return to the Euler equation,

$$x'' = g(x), \qquad x(0) = c, \qquad x'(T) = 0. \tag{5.23.1}$$

In place of studying this equation directly, we pose the problem of minimizing some appropriate norm $\|x'' - g(x)\|$, where x is subject to the same end point constraints. The simplest and most frequently employed measure is

$$J(x) = \int_0^T (x'' - g(x), x'' - g(x)) \, dt. \tag{5.23.2}$$

Once we have transformed the problem into that of minimizing $J(x)$, we proceed to introduce a trial function depending only of a finite number of parameters. In general, the parameters are introduced in a linear fashion,

$$x \cong \sum_{k=1}^{M} a_k \, y^{(k)}, \tag{5.23.3}$$

where the $y^{(k)}$ are specified functions. The problem of minimizing $J(x)$ has been reduced to a finite-dimensional problem. The method is both powerful and versatile as results in the references at the end of this chapter will demonstrate. Let us note that we can continue in the same vein and replace $J(x)$ by $J(x) + \lambda \, \|x'(T)\|$, $\lambda \gg 1$, if for some reason it is inconvenient to impose the condition $x'(T) = 0$.

To validate the method, stability theory is required. We must demonstrate that under appropriate conditions a function satisfying the condition $\|y'' - g(y)\| \leq \varepsilon$, $y(0) = c$, $y'(T) = 0$, is close to the desired function x.

EXERCISES

1. Applying the foregoing method to the solution of $Ax = b$.
2. Apply the method to the solution of $x'' - Ax = 0$, $x(0) = c$, $x'(T) = 0$.
3. Discuss the use of the foregoing method for suboptimization purposes.

5.24.   Method of Moments

In place of the foregoing method it is sometimes convenient to proceed as follows. Let $\{\phi_k(t)\}$, $k = 1, 2, \ldots, M$, be a given sequence of functions and consider the set of M "moment equations"

$$\int_0^T (x'' - g(x), \phi_k(t)) \, dt = 0, \qquad k = 1, 2, \ldots, M. \tag{5.24.1}$$

Taking x to have the form given in (5.23.3), we obtain a set of M simultaneous equations for the parameters $a_1, a_2, \ldots, a_M$.

If $g(x)$ is a linear operator, the resulting equations are linear. If not, various additional techniques must be employed to obtain a numerical solution.

EXERCISES

1. For the case where $g(x)$ is linear, what is the connection between the Bubnov-Galerkin method and the moment method?
2. Apply this method to $x'' - Ax = 0$, $x(0) = c$, $x'(T) = 0$. How might we go about choosing the $\phi_k(t)$?

5.25.　Gradient Methods

An equation of the form

$$N(x) = 0 \tag{5.25.1}$$

can always be considered to be the steady-state form of a dynamic equation such as

$$x_{n+1} - x_n = N(x_n), \qquad x_0 = c, \tag{5.25.2}$$

or, more generally,

$$x_{n+1} - x_n = \phi(N(x_n)), \qquad x_0 = c. \tag{5.25.3}$$

Here, ϕ is a function chosen to ensure stability and rapidity of convergence, and such that $\phi(N) = 0$ implies $N = 0$. This is the basic idea of the method of successive approximations.

A continuous version of (5.25.2) is the partial differential equation

$$\frac{\partial y}{\partial t} = N(y), \tag{5.25.4}$$

and a continuous version of (5.25.3) is

$$\frac{\partial y}{\partial t} = \phi(N(y)). \tag{5.25.5}$$

The basic idea is to prescribe an initial condition

$$y_{t=0} = z. \tag{5.25.6}$$

and then to integrate (5.25.4) numerically as an *initial-value* problem. If carried out adroitly, the limit of y as $t \to \infty$ will exist and furnish a solution of (5.25.1).

As to be expected, a certain amount of preliminary analysis is required to ensure the success of the method. There are questions of both analytic and computational stability as well as questions of time and storage.

EXERCISES

1. When can we obtain the solution of $Ax = b$ as the limit of $x' = Ax - b$?
2. What about the equation $x' = -(A^T Ax - x^T b)$ as a means of obtaining the solution of $Ax = b$? Has it any advantage over the simpler equation in Exercise 1?

5.26. A Specific Example

To illustrate the kinds of questions that arise in the course of applying this technique, let us return to the equation considered above,

$$u'' = g(u), \qquad u(0) = c, \qquad u'(T) = 0. \tag{5.26.1}$$

Proceeding rather directly, this leads to the partial differential equation

$$v_t = v_{ss} - g(v),$$

$$v(0, t) = c, \qquad v_s(T, t) = 0, \qquad t > 0, \tag{5.26.2}$$

$$v(s, 0) = k(s), \qquad 0 \le s \le T,$$

where $k(s)$ remains to be chosen. We have replaced the former independent variable by s in order to use t for the new gradient variable; see Figure 5.1.

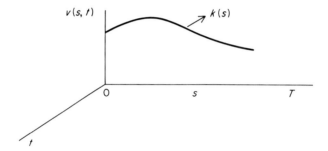

Figure 5.1

Let us restrict ourselves to a quick examination of the question of stability of the limit function. Set

$$v(s, t) = u(s) + w(s, t), \tag{5.26.3}$$

where u is a solution of (5.26.1). Then, substituting in (5.26.2),

$$w_t = u_{ss} + w_{ss} - g(u + w) = [u_{ss} - g(u)] + [w_{ss} - wg_u] + \cdots. \tag{5.26.4}$$

Consider the first order approximation, the linear partial differential equation

$$w_t = w_{ss} - wg_u, \tag{5.26.5}$$

subject to the initial and boundary conditions

$$w(s, 0) = k(s) - u(s), \tag{5.26.6}$$

$$v(0, t) = v_s(T, t) = 0.$$

We see that the asymptotic behavior of w depends upon the location of the largest characteristic value of the Sturm-Liouville problem

$$\lambda z = z'' - zg_u,$$

$$z(0) = 0, \qquad z'(T) = 0. \tag{5.26.7}$$

If all of the characteristic values are negative, we can assert that the gradient method is valid, provided that $\| u - k \|$ is sufficiently small. In other words, as usual in the solution of nonlinear problems, a method of successive approximations is effective if we start sufficiently close to the solution. This is again a result in stability theory.

EXERCISES

1. Examine the application of the method to the solution of $u'' - au = 0$, $u(0) = c$, $u'(T) = 0$, and to the solution of $u'' + au = 0$, $u(0) = c$, $u'(T) = 0$, where in both cases $a > 0$.
2. Examine the solution of the scalar equation $g(r) = 0$ by means of the differential equation $u' = g(u)$, $u(0) = c$.
3. Can we modify (5.26.2) in such a way that the perturbation equation in (5.26.7) possesses only characteristic roots which are negative?
4. How might we go about choosing an initial approximation $k(s)$?

5.27. Numerical Procedure: Semi-discretization

A direct approach along the foregoing lines involves the numerical solution of the partial differential equation of (5.25.2), continuing until the steady-state behavior takes over. This can be made equivalent to the task of numerical integration of a large system of ordinary differential equations. For example, we can employ a semi-discretization procedure, writing

$$u'_n = \frac{u_{n+1} + u_{n-1} - 2u_n}{\Delta^2} + g(u_n), \tag{5.27.1}$$

$$u_n(0) = k_n, \qquad n = 1, 2, \ldots, N - 1,$$

where $N\Delta = 1$ and

$$u_0(t) = c, \qquad u_N(t) - u_{N-1}(t) = 0. \tag{5.26.2}$$

Here $u_n(t)$ is an approximation to the desired value $u(n\Delta, t)$.

The time required for convergence depends critically upon the discretization step Δ and the choice of the initial approximation, $k(s)$. Let us note that any of the methods previously discussed can be used to furnish a suitable initial approximation. This is a good example of the utility of combining several different methods.

5.28. Nonlinear Extrapolation

Our aim is to estimate $u(s)$, $0 \le s \le T$ given values of $v(s, t)$. One way of doing this, as indicated above, is to use the relation

$$u(s) \cong v(s, t) \tag{5.28.1}$$

for $t \gg 1$. We would expect, however, that we could obtain a far more accurate value for $u(s)$ by using the more precise asymptotic behavior

$$v(s, t) \cong u(s) + e^{-\lambda t} u_1(s), \tag{5.28.2}$$

obtained from Section 5.26. Using this relation, we can obtain accurate values of $u(s)$ from values of $v(s, t)$ for relatively small values of t. This means both a considerable decrease in computing time and also an avoidance of various troublesome questions that arise from numerical integration of differential systems over long time intervals.

Let us take the problem in the following discrete form: Given the relation

$$r_n = a_1 + b_1 \lambda^n, \quad |\lambda| < 1, \tag{5.28.3}$$

obtain an estimate for a_1 in terms of values of r_n.

A direct process of elimination yields the expression

$$a_1 = \frac{(r_n r_{n+2} - r_{n+1}^2)}{(r_n + r_{n+2} - 2r_{n+1})}. \tag{5.28.4}$$

This can be looked upon as a "nonlinear filter" for the term $b\lambda_1{}^n$. Thus, if r_n actually possesses the asymptotic expression

$$r_n = a_1 + b_1 \lambda_1{}^n + b_2 \lambda_2{}^n + \cdots, \tag{5.28.5}$$

with $1 > |\lambda_1| > |\lambda_2| > \cdots$, a direct calculation shows that

$$s_n = \frac{(r_n r_{n+2} - r_{n+1}^2)}{(r_n + r_{n+2} - 2r_{n+1})} = a + 0(|\lambda_2|^n). \tag{5.28.6}$$

On the other hand, from (5.28.5),

$$r_n = a + 0(|\lambda_1|^n). \tag{5.28.7}$$

Thus, if $|\lambda_2| \ll |\lambda_1|$, the use of s_n can result in a tremendous gain in accuracy. Examples of this will be found in the references at the end of the chapter.

EXERCISES

1. What are continuous versions of the foregoing results?
2. What does one do in the vector case?

5.29.   Rayleigh-Ritz Methods

Let us turn finally to a brief description of one of the most powerful methods of mathematical physics and applied mathematics—the celebrated technique of Rayleigh-Ritz.

Recall that our original aim was to resolve a certain minimization problem' that of minimizing the functional

$$J(x) = \int_0^T [(x', x') + 2h(x)] \, dt, \qquad (5.29.1)$$

subject to $x(0) = c$. This leads to the problem of solving the Euler equation

$$x'' - \text{grad } h(x) = 0, \qquad x(0) = c, \qquad x'(T) = 0. \qquad (5.29.2)$$

Keeping in mind Jacobi's dictum*, why not reverse the process and study the solution of (5.29.2) in terms of the minimization of $J(x)$. This is the essence of the celebrated Rayleigh-Ritz method. We can now proceed as in Section 5.23.

EXERCISES

1.   Apply the preceding method to the functional

$$J(x) = \int_0^T [(x', x') + (x, A(t)x)] \, dt,$$

where $x(0) = c$.

2.   Apply the method to the linear algebraic equation $Ax = b$.

3.   The minimizing parameters a_i are functions of M, $a_i(M)$, when we set $x(t) = \sum_{i=1}^M a_i y_i(t)$. Discuss the possibility of using extrapolation techniques to determine $a_i(\infty)$.

Miscellaneous Exercises

1.   Obtain the Euler equation for the minimization of $J(u) = \int_0^T g(u'(t), u(t), u(t - 1)) \, dt$, $u(t) = h(t)$, $-1 \le t \le 0$.

2.   What are some of the difficulties that confront a numerical solution?

3.   Let $E(u, u', u'') = 0$ be the Euler equation associated with $J(u) = \int_0^T g(u, u') \, dt$. Are there any advantages to replacing the original variational problem by the problem of minimizing $\int_0^T [g(u, u') + \lambda E(u, u', u'')^2] \, dt$? (See

A. K. Rigler, "The Construction of Difference Approximations From A "Sensitized Functional," *Quart. Appl. Math.*, **XXVI**, 1968, pp. 288–290.)

* *Mann immer umkehren muss!*—Jacobi.

BIBLIOGRAPHY AND COMMENTS

5.1.   An important survey of current techniques for treating two-point boundary-value problems is

R. Conti, "Recent Trends in the Theory of Boundary Value Problems for Ordinary Differential Equations," *Boll. Unione Mat. Ital.*, **22** (3), 1967, pp. 135–178.

For a discussion of computational methods see

L. Collatz, *The Numerical Treatment of Differential Equations*, Springer, Berlin, 1966.
5.3.   See

R. Bellman, "Successive Approximations and Computer Storage Problems in Ordinary Differential Equations," *Comm. Assoc. Comput. Machinery*, **4**, 1961, pp. 222–223.
R. Bellman and R. Kalaba, *Quasilinearization and Nonlinear Boundary-value Problems*, Elsevier, New York, 1965.
5.4.   See

R. Bellman, "On the Computational Solution of Differential-difference Equations," *J. Math. Anal. Appl.*, **2**, 1961, pp. 108–110.
R. Bellman, "From Chemotherapy to Computers to Trajectories," *Math. Prob. Biol. Sci.* American Mathematical Society, 1962, pp. 225–232.
R. Bellman and K. L. Cooke, "On the Computational Solution of a Class of Functional Differential Equations," *J. Math. Anal. Appl.*, **12**, 1965, pp. 495,–500.
5.5.   See

R. Bellman, "Dynamic Programming, Generalized States, and Switching Systems," *J. Math. Anal. Appl.*, **12**, 1965, pp. 360–363.
R. Bellman, "A Function is a Mapping—plus a Class of Algorithms," *Inform. Sci.*, to appear.

See also

M. Aoki, "On Optimal and Suboptimal Policies in Control Systems," Chapt. 1 in *Advances in Control Systems*, Vol. I; *Theory and Applications*, (C. T. Leondes, ed.), Academic Press, New York, 1964.
S. Dreyfus, "The Numerical Solution of Variational Problems," *J. Math. Anal. Appl.*, **5**, 1962, pp. 30–45.
J. K. Hale and J. P. LaSalle, "Differential Equations: Linearity vs. Nonlinearity," *SIAM Rev.* **5**, 1963, pp. 249–272.
M. Sobral, "Sensitivity in Optimal Control Systems," *Proc. IEEE*, **56**, 1968, pp. 1644–1752. (This contains an excellent bibliography.)
J. H. Westcott, J. J. Florentin and J. D. Pearson, "Approximation Methods in Optimal and Adaptive Control, *Proc. IFAC Sec. Congr.*, Basel, 1963.

and the book

P. B. Bailey and L. Shampine, *Nonlinear Two-point Boundary-value Problems*, Academic Press, New York, 1969.

5.6.   See
R. Bellman and R. Kalaba, *Quasilinearization and Nonlinear Boundary-value Problems*, Elsevier, New York, 1965.

For a use of a second-order approximation together with dynamic programming, see

S. E. Dreyfus, "The Numerical Solution of Nonlinear Optimal Control Problems," *Numerical Solutions of Nonlinear Differential Equations*, Wiley, New York, 1966, pp. 97–113.
5.7.   See

R. Bellman, H. Kagwada and R. Kalaba, "Nonlinear Extrapolation and Two-point Boundary-value Problems," *Comm. A.C.M.*, **8**, 1965, pp. 511–512.
5.9.   For an interesting use of quadratic convergence, see

J. Moser, "A New Technique for the Construction of Solutions of Nonlinear Differential Equations," *Proc. Nat. Acad. Sci. USA*, **47**, 1961, pp. 1824–1831.
5.10.   See the paper and Bellman-Kalaba book referred to in 5.3 for the details.
5.12.   See

R. Bellman, *Introduction to the Mathematical Theory of Control Processes*, Vol. I: *Linear Equations and Quadratic Criteria*, Academic Press, New York, 1968.
5.14.   For a discussion of stability and instability, see

R. Bellman, *Methods of Nonlinear Analysis*, Academic Press, New York, 1970.

It is essential when discussing the properties of the solution of an equation to distinguish between its stability as a function of the initial conditions, length of the interval and structure of the equation, and the stability of a particular computational procedure used to obtain the solution. Thus, for example, the solution of $u'' - u = 0$, $u(0) = c$, $u'(T) = 0$ is stable as far as changes in c and T are concerned and, as we shall see in Chapter 8, as far as the addition of nonlinear terms is concerned. But, it is not stable as far as determination by solution of the associated initial value problem is concerned. Quadrature techniques, however, can be used to provide stable computational schemes. I am grateful to G. Dahlquist for clarification of these points.
5.16.   See

G. E. Forsythe, "Today's Computational Methods of Linear Algebra," *SIAM Rev.*, **9**, 1967, pp. 489–515.
R. Bellman, *Introduction to Matrix Analysis*, 2nd ed. McGraw-Hill, New York, 1970.
5.17.   See the above cited book on matrix analysis, and

R. Bellman, R. Kalaba, and J. Lockett, *Numerical Inversion of the Laplace Transform*, Elsevier, New York, 1966.
R. Lattes, and J. L. Lions, *The Method of Quasi-reversibility: Applications to Partial Differential Equations*, Elsevier, New York, 1969.

For entirely different methods based upon stochastic processes, see

J. N. Franklin, *Well-posed Stochastic Extensions of Ill-Posed Linear Problems*, Tech. Rp. 135, California Institute of Technology, Pasadena, 1969.
5.18.   See

R. Bellman, R. Kalaba, and J. Lockett, *Numerical Inversion of the Laplace Transform*, Elsevier, New York, 1966.

for an introductory account of quadrature, and

D. E. Greenspan, "Approximate Solution of Initial Value Problems for Ordinary Differential Equations by Boundary-value Techniques," *J. Math. Phys. Sci.*, **1**, 1967, pp. 261–274.

R. Bellman and J. Casti, "Differential Quadrature and Long-Term Integration," *JMAA*, to appear.
  5.20. See

S. R. McReynolds and A. E. Bryson, "A Successive Sweep Method for Solving Optimal Programming Problems," *Joint Automat. Control Conf.*, Troy, New York, 1965.
  5.23–24. See the book cited in Section 5.14.

For some discussions of suboptimization, see

D. L. Kleinman, T. Fortmann and M. Athans, "On the Design of Linear Systems with Piecewise-Constant Feedback Gains," *IEEE, Trans. Automat. Control* **AC-13**, pp. 354–361.
P. V. Kokotovic and P. Sannuti, "Singular Perturbation Method for Reducing the Model Order in Optimal Control Design," *IEEE Trans. Automat. Control*, **AC-13**, 1968, pp. 377–383.
  5.25. See

P. Rosenbloom, "The Method of Steepest Descent," *Proc. Symp. Appl. Math.* **VI**, 1956, pp. 127–176.
  5.28. See

R. Bellman, and R. Kalaba "A Note on Nonlinear Summability Techniques in Invariant Imbedding," *J. Math. Anal. Appl.*, **6**, 1963, pp. 465–472.
  5.29. See the book cited in Section 5.14 and, for example,

S. Altshuler, "Variational Principles for the Wave Equation Function in Scattering Theory," *Phys. Rev.*, **109**, 1958, pp. 1830–1836.
T. Ikebe and T. Kato, "Applications of Variational Methods to the Thomas-Fermi Equation," *J. Phys. Soc. Japan*, **12**, 1957, pp. 201–203.

6

CONTINUOUS CONTROL PROCESSES
AND DYNAMIC PROGRAMMING

6.1. Introduction

In Chapters 2 and 3 we studied various analytic and computational aspects of discrete control processes using dynamic programming; in Chapters 4 and 5 we examined corresponding aspects of continuous control processes relying on the calculus of variations. Let us now blend a number of different ideas and examine the application of dynamic programming to continuous control processes. The basic theme is a simple one: a continuous control process can be considered to be a multistage decision process of continuous type.

On one hand, a formulation in terms of continuous time simplifies the analysis considerably. The calculus of derivatives is much simpler than the calculus of finite differences; integration is far simpler than summation. On the other hand, the transcendental operations of calculus introduce certain technical difficulties which are not present in the case of discrete processes. Questions concerning the existence of derivatives now become vital to a rigorous foundation.

The basic problem we face in this chapter is to present a number of fundamental ideas without becoming deeply enmeshed in distracting technicalities, which are either extraneous to the central themes of control theory or of lesser import. In some cases surmounting the analytic detail becomes a major task. Whenever we feel that providing all of the filigree becomes too onerous, we shall refer to various research papers.

6.2. Continuous Multistage Decision Processes

Let us begin with the familiar problem of minimizing the scalar functional

$$J(u) = \int_0^T g(u, u') \, dt, \qquad (6.2.1)$$

over the class of functions for which the integral exists and which are subject to the condition $u(0) = c$. In pursuing the route of the calculus of variations, the function $u(t)$ is regarded as a single entity, a point in function space. We perturb in the neighborhood of this "point" in order to obtain the basic necessary condition, the Euler equation; see Figure 6.1.

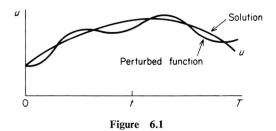

Figure 6.1

The curve itself is considered to be a locus of points, profitably thought of as a trajectory over time—a path traced out by a particle moving according to certain specified laws.

To apply the methods of dynamic programming we insert ourselves into a different conceptual framework. The minimization of $J(u)$ is regarded as the task of determining an optimal policy or, more generally, all optimal policies. In this case, this involves ascertaining the slope u' at each point on a solution curve as a function of the coordinates of the point, which is to say, $u' = k(u, t)$; see Figure 6.2.

The curve itself is considered to be an envelope of tangents. We can visualize the function u, the trajectory over time, as the result of "steering" a point vehicle; u' determines the direction at each point.

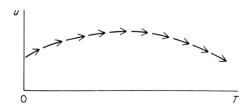

Figure 6.2

6.3.   Duality

That the two viewpoints, so different in concept, are mathematically equivalent is a consequence of the fundamental duality of Euclidean geometry. This
fortuitous and remarkable equivalence holds, however, only for deterministic
control processes. As soon as we allow stochastic effects in the system, we
perceive an enormous difference between these two formulations. These fascinating and important problems will be discussed in Volume III. Here we
consider only deterministic control processes.

In studying any particular deterministic control process we have a choice
of which approach to employ, that of the calculus of variations or that of
dynamic programming. As to be expected, some combination of the two dual
approaches will generally be more powerful than either of the two procedures
alone. What combination to use depends upon the particular control process.

We shall return to this duality in Chapter 9 where the basic idea is utilized
to obtain upper and lower bounds for the value of the minimum of $J(u)$.

6.4.   Analytic Formalism

Let us see the nature of results that can be obtained from the dynamic programming approach. To begin with, we will proceed in a purely formal
fashion. After we have convinced ourselves that there is perhaps something
of merit to be gained, we will provide a rigorous basis for the results derived
in a heuristic fashion.

Corresponding to our approach in the discrete case, we begin by introducing
the basic function

$$f(c, T) = \min_{u} J(u) = \min_{u} \int_0^T g(u, u') \, dt, \qquad (6.4.1)$$

the return function. This function plays a paramount role in all that follows.
The quantity c is the initial value, $u(0)$; f is initially defined for $T \geq 0$,
$-\infty < c < \infty$.

To derive an equation for $f(c, T)$, we decompose the interval $[0, T]$ into
two sub-intervals $[0, \Delta]$ and $[\Delta, T]$; see Figure 6.3.

Figure   6.3

Use the additivity of the integral to write

$$\int_0^T = \int_0^\Delta + \int_\Delta^T, \qquad (6.4.2)$$

and observe that

$$f(c, T) = \min_{u[0, T]} = \min_{u[0, \Delta]} \min_{u[\Delta, T]}. \qquad (6.4.3)$$

Assuming for the moment that u is a well-behaved function, the choice of a function over $[0, \Delta]$ can be considered to be a choice of an initial slope, $u'(0)$, provided that $\Delta \ll 1$. This is to say, we approximate to $u(t)$ over $[0, \Delta]$ by a straight line. Let us set

$$v = u'(0). \qquad (6.4.4)$$

The quantity v depends on c, the initial state and T, the duration of the process; $v = v(c, T)$.

The value of u at Δ is then, to terms which are $o(\Delta)$,

$$u(\Delta) = c + v\Delta. \qquad (6.4.5)$$

To the same degree of approximation,

$$\int_0^\Delta g(u, u') \, dt = g(c, v) \, \Delta. \qquad (6.4.6)$$

The determination of the portion of the minimizing function over $[\Delta, T]$ may be regarded as a process of exactly the same type as the original, taking into account the transformation

$$c \rightarrow c + v\Delta,$$
$$T \rightarrow T - \Delta. \qquad (6.4.7)$$

Hence, by definition of f,

$$\min_{u[\Delta, T]} \int_\Delta^T = f(c + v \Delta, T - \Delta), \qquad (6.4.8)$$

again to terms which are $o(\Delta)$. Combining (6.4.3), (6.4.6), and (6.4.8), and invoking the principle of optimality, we obtain the approximate functional equation

$$f(c, T) = \min_v [g(c, v)\Delta + f(c + v\Delta, T - \Delta)] + o(\Delta), \qquad (6.4.9)$$

valid for $\Delta \ll 1$, assuming as we do that all of our operations are legal. There are clearly some difficulties in justifying the $o(\Delta)$ term in (6.4.9).

6.5. Limiting Form

To obtain the limiting form of this equation as $\Delta \to 0$, we write

$$f(c + v\Delta, T - \Delta) = f(c, T) + (vf_c - f_T)\Delta + o(\Delta), \qquad (6.5.1)$$

assuming the existence of the partial derivatives appearing. The relation in (6.4.9) then yields in the limit as $\Delta \to 0$, the equation

$$f_T = \min_v [g(c, v) + vf_c]. \qquad (6.5.2)$$

The initial condition is readily seen to be

$$f(c, 0) = 0. \qquad (6.5.3)$$

Observe, as previously remarked, that the v which minimizes in (6.5.2) is a function of c and T. We call this function, $v(c, T)$, the *policy function*.

It remains to be seen whether (6.5.2), a nonlinear partial differential equation of novel form, can be used in any fruitful way to treat continuous control processes.

EXERCISE

1. Obtain the corresponding equation for the case where

$$J(u, v) = \int_0^T g(u, v)\, dt, \ u' = h(u, v), \ u(0) = c.$$

6.6. Discussion

As in the case of discrete processes, a single dynamic programming equation determines *two* functions: the return function $f(c, T)$ and the policy function $v(c, T)$. Again as in the discrete case, a knowledge of one can be used to determine the other. This double-barrelled approach supplies a flexibility which can be used to good avail.

6.7. Associated Nonlinear Partial Differential Equations

From (6.5.2) we can readily derive a pair of equations for both f and v. Determining the location of the minimizing value by means of differentiation (we continue to proceed formally and hope for the best), we obtain the equation

$$0 = g_v + f_c. \qquad (6.7.1.)$$

Combining this with the original equation, we are led to the simultaneous equations

$$f_T = g + vf_c,$$
$$f_c = -g_v,$$
(6.7.2)

for the functions f and v.

This system can be put into the more symmetric form

$$f_T = g - vg_v,$$
$$f_c = -g_v.$$
(6.7.3)

If we regard (6.7.1) as an equation which determines v in terms of f_c, namely

$$v = h(c, f_c),$$
(6.7.4)

we can write the first equation in (6.7.2) in the form

$$f_T = \phi(c, f_c), \qquad f(c, 0) = 0,$$
(6.7.5)

where the function ϕ is determined by (6.7.3) and (6.7.4). It is plausible that this equation determines $f(c, T)$ as the solution of an initial-value problem.

It is equally easy to eliminate f between the two equations of (6.7.3) by simple differentiation. We have

$$(f_T)_c = (g - vg)_c = g_c - vg_{vc} - v_c g_{vv},$$
$$(f_c)_T = -g_{vv} v_T.$$
(6.7.6)

Equating f_{cT} and f_{Tc}, we have an equation for v,

$$g_c - vg_{vc} - vv_c g_{vv} = -g_{vv} v_T.$$
(6.7.7)

The initial condition is

$$v(c, 0) = r(c),$$
(6.7.8)

where $r(c)$ is the value of v which minimizes $g(c, v)$.

Observe that the equation for f is nonlinear, while the equation for v is quasilinear. By this we mean that the equation for v is linear in v_c and v_T.

EXERCISES

1. Write the equations for f and v when $g(u, u') = u'^2 + h(u)$.
2. Write the equations for f and v when $g(u, u') = u'^2 + u^2$.
3. From (6.7.5) obtain a quasilinear equation for f_c. What is the meaning of f_c?

6.8. Characteristics and the Euler Equation

It is natural to seek some links between the foregoing analytic formulation of the determination of the optimal policy, and thus the minimizing function, and the classical Euler equation. The crucial observation is that (6.7.7) is a quasilinear partial differential equation which possesses *characteristics* taking the form of ordinary differential equations.

Since it is intuitively clear that there is only one set of ordinary differential equations associated with the original minimization problem (the minimization of

$$J(u) = \int_0^T g(u, u') \, dt, \tag{6.8.1}$$

we anticipate that the equations for the characteristics must be equivalent to the equation

$$d/dt(g_{u'}) = g_u. \tag{6.8.2}$$

We can also anticipate this from geometric considerations. The duality between the calculus of variations and dynamic programming must express itself in the relation between the equation for $v(c, T)$ and the characteristic equations.

EXERCISE

1. Derive the Euler equation directly from (6.5.2) without use of the theory of characteristics. (See Dreyfus, 1965.)

6.9. Rigorous Aspects

Let us now turn to the question of providing a rigorous basis for the fundamental equation of (6.5.2). There are two approaches we can use. To begin with, we can provide a direct verification based on the results of Chapter 4. We considered the particular case

$$g(u, u') = u'^2 + g(u) \tag{6.9.1}$$

in Sections 4.19–4.20. We urge the reader to complete the details for the more general case where $g(c, v)$ is convex in v for all c, and for the multidimensional case as well.

Alternatively, armed with information concerning properties of the minimizing function $u(t)$ or, equivalently, the nature of the optimal policy, we can provide a direct proof. As is to be expected, a proof will provide only local validity, in some region $|c - c_0| \ll 1$, $T \ll 1$. Call this region R.

We begin with the fact that $f(c, T)$ exists in a region $|c - c_0| \ll 1$, $T \ll 1$ and possesses in this region continuous partial derivatives with respect to c and T, provided that g is a "reasonable" function. We shall assume that $g(u, u')$ is such that $g(c, v)$ is uniformly convex in v for $|c - c_0| \ll 1$. From this we can conclude, using the methods of Chapter 4, that u possesses both first and second derivatives for $0 \le t \le T$. Hence, for the minimizing function we have the estimate

$$u = c + vt + 0(t^2), \qquad (6.9.2)$$

for $0 \le t \le T$, with the 0-term *uniform* in the region R. Furthermore, we know that $|v| \le v_0$ in this region.

Thus, returning to the relation defining f,

$$f(c, T) = \min_{u[0, \Delta]} \min_{u[\Delta, T]} \left[\int_0^\Delta + \int_\Delta^T \right], \qquad (6.9.3)$$

we write

$$f(c, T) = \min_{u[0, \Delta]} \left[\int_0^\Delta + \min_{u[\Delta, T]} \int_\Delta^T \right], \qquad (6.9.4)$$

whence

$$f(c, T) = \min_{u[0, \Delta]} \left[\int_0^\Delta g(u, u') \, dt + f(u(\Delta), T - \Delta) \right]. \qquad (6.9.5)$$

We know that the function $f(u(\Delta), T - \Delta)$ is well-defined since $|u(\Delta) - c| \ll 1$ for Δ small, by virtue of our a priori bound on v.

Using our knowledge of the behavior of u and f, we replace (6.9.5) by

$$f(c, T) = \min_{u[0, \Delta]} [g(c, v)\Delta + o(\Delta) + f(c + v\Delta, T - \Delta)] + o(\Delta), \quad (6.9.6)$$

where now the o-terms are uniform in R. Similarly, the minimum over functions $u(t)$ defined over $[0, \Delta]$ can be replaced by

$$f(c, T) = \min_{|v| \le v_0} [g(c, v)\Delta + f(c + v\Delta, T - \Delta)] + o(\Delta). \qquad (6.9.7)$$

Applying Taylor's theorem, we have

$$f(c + v\Delta, T - \Delta) = f(c, T) + (vf_c - f_T)\Delta + o(\Delta), \qquad (6.9.8)$$

again with a uniform o-term for $|v| \le v_0$, which as we know holds for $|c - c_0| \ll 1$, $T \ll 1$. Thus,

$$0 = \min_{|v| \le v_0} [g(c, v) + (vf_c - f_T)] + o(1), \qquad (6.9.9)$$

whence, letting $\Delta \to 0$, we have the result

$$0 = \min_{|v| \le v_0} [g(c, v) + vf_c - f_T].$$
(6.9.10)

Finally, since the minimand is convex and we know that the absolute minimum occurs in the region $|v| \le v_0$, we can remove the constraint and write

$$0 = \min_{v} [g(c, v) + vf_c - f_T],$$
(6.9.11)

the desired equation.

6.10. Multidimensional Case

There is no particular difficulty involved in extending these results to the multidimensional case. Consider the problem of minimizing

$$J(x) = \int_0^T g(x, x') \, dt,$$
(6.10.1)

where x is an N-dimensional vector subject to $x(0) = c$. Setting

$$f(c, T) = \min_{x} J(x),$$
(6.10.2)

we readily obtain the nonlinear partial differential equation

$$f_T = \min_{y} [g(c, y) + (\operatorname{grad} f, y)],$$
(6.10.3)

with the initial condition

$$f(c, 0) = 0.$$
(6.10.4)

A rigorous derivation in a region $\|c - c_0\| \ll 1, T \ll 1$ follows the preceding lines, under the assumption that g is uniformly convex in x'.

6.11. Riccati Equation

A case of particular importance, both intrinsically and as a foundation stone for the utilization of the method of successive approximations, is that where $g(x, x')$ is quadratic in x and x'. To illustrate, consider the question of minimizing

$$J(x) = \int_0^T [(x', x') + (x, Ax)] \, dt,$$
(6.11.1)

where $x(0) = c$ and A is positive definite.

The equation derived from (6.10.3) is

$$f_T = \min_y \; [(y, y) + (\operatorname{grad} f, y)]. \qquad (6.11.2)$$

Using the fact that $f(c, T)$ is quadratic in c,

$$f(c, T) = (c, R(T)c), \qquad (6.11.3)$$

we can reduce (6.11.2) to an ordinary differential equation.

Substituting (6.11.3) in (6.11.2), we find that R satisfies the Riccati differential equation

$$R' = A - R^2, \qquad R(0) = 0. \qquad (6.11.4)$$

The optimal policy is given by

$$y = -R(T)c. \qquad (6.11.5)$$

We have discussed these matters in some detail in Volume I, with particular emphasis upon the serious computational difficulties encountered when the dimension of R is large.

EXERCISES

1. Find the Riccati equations associated with the cases

(a) $\quad J(x, y) = \displaystyle\int_0^T [(x, x) + (y, y)] \, dt \qquad x' = Ax + By, \qquad x(0) = c;$

(b) $\quad J(x, y) = (x(T), Cx(T)) + \displaystyle\int_0^T (y, y) \, dt, \; x' = Ax + By, \qquad x(0) = c.$

2. Show that the partial differential equation of (6.9.11) can be reduced to an ordinary differential equation if

$$J(u) = \int_0^T (u^4 + u'^4) \, dt \qquad u(0) = c.$$

Does the differential equation hold for all $T \geq 0$?

3. Generally, consider the same questions for the minimization of

$$J_n(u) = \int_0^T (u^{2n} + (u')^{2n}) \, dt, \qquad u(0) = c.$$

4. Does $\lim_{n \to \infty} [\min_u J_n(u)]^{1/n}$ exist? Does the corresponding Euler equation approach a limit? Is there any connection between the limit of the solution and the solution of the limiting problem? (*Hint:*

$$\lim_{n = \infty} \left(\int_0^T f^{2n} \, dt \right)^{1/2n} = \max_{0 \leq t \leq T} |f|.$$

See the references cited at the end of Chapter 4, Section 24.)

6.12. Computational Significance

We have shown that under suitable conditions the problem of determining the minimum of

$$J(u) = \int_0^T g(u, u') \, dt, \tag{6.12.1}$$

subject to $u(0) = c$, can be transformed into the problem of solving the non-linear partial differential equation

$$f_T = \min_v [g(c, v) + vf_c], \qquad f(c, 0) = 0. \tag{6.12.2}$$

One of these assumptions is that g is uniformly convex in v, which means that we can solve for the minimizing v in terms of f_c and write (6.12.2) in the form

$$f_T = h(c, f_c), \qquad f(c, 0) = 0. \tag{6.12.3}$$

This is an initial-value problem.

Can we use this approach to obtain a numerical solution of the original variational problem? There are a number of powerful approaches we can contemplate based upon the use of (6.12.3). We shall discuss some of these below.

6.13. Finite Difference Techniques

If we employ conventional finite difference techniques to obtain the numerical solution of an equation such as (6.12.3), we can expect some difficulties. Consider, for example, an equation such as

$$f_T = g(c) + f_c^2, \qquad f(c, 0) = 0. \tag{6.13.1}$$

We can readily construct a computational scheme such as

$$\frac{\phi(c, T + \Delta) - \phi(c, T - \Delta)}{2\Delta} = g(c) + \frac{[\phi(c + \delta, T) - \phi(c - \delta, T)]^2}{4\delta^2}, \tag{6.13.2}$$

where $T = \Delta, 2\Delta, 3\Delta, \ldots,$ $c = 0, \pm\delta, \pm2\delta, \ldots, \pm M\delta$, and, hopefully, $\phi(c, T) \cong f(c, T)$. There is the usual difficulty of an expanding grid which we will ignore. Let us write (6.13.2) in the form

$$\phi(c, T + \Delta) = \phi(c, T - \Delta) + 2\Delta g(c) + (\Delta/2\delta^2)[\phi(c + \delta, T) - \phi(c - \delta, T)]^2. \tag{6.13.3}$$

The danger inherent in a scheme of this nature is clear. We are multiplying a small quantity $[\phi(c + \delta, T) - \phi(c - \delta, T)]^2$ by a quantity which may be. large, $\Delta/2\delta^2$. To avoid this difficulty, we want to take $\Delta/2\delta^2$ of moderate size If we do so, however, the stepsize in time, Δ, will be of the order of magnitude of δ^2, the square of the stepsize in space. This will entail a large amount of computing time if δ is taken small to ensure accuracy.

As an alternate to this approach, we can convert an equation of the form

$$f_T = h(c, f_c), \qquad f(c, 0) = r(c), \tag{6.13.4}$$

into an equation of quasilinear type and then proceed as we discussed in Section 6.12. This is readily done by partial differentiation. We have

$$(f_c)_T = h_1(c, f_c) + h_2(c, f_c)(f_c)_c, \tag{6.13.5}$$

where

$$h_1 = \frac{\partial h}{\partial c}, \qquad h_2 = \frac{\partial h}{\partial f_c}, \tag{6.13.6}$$

Setting $f_c = w$, we obtain a quasilinear equation

$$w_T = h_1(c, w) + h_2(c, w)w_c, \qquad w(c, 0) = r_c. \tag{6.13.7}$$

We recognize f_c as the marginal return function. A scheme analogous to (6.13.3) produces the ratio Δ/δ, allowing a much larger step in time.

Once the foregoing equation has been derived, the theory of characteristics can be employed to provide an alternate computational algorithm.

EXERCISE

1. We encounter the same problem in solving $u_t = u_{xx}$ using the recurrence relation

$$\frac{u(x, t + \Delta) - u(x, t - \Delta)}{2\Delta} = \frac{u(x + \delta, t) + u(x - \delta, t) - 2u(x, t)}{\delta^2}.$$

Show that we can avoid this difficulty here by using suitable relations of the form

$$v(x, t + \Delta) = \sum_{i=1}^{R} a_i[v(x - \delta_i, t) + v(x + \delta_i, t)],$$

together with interpolation, where $a_i \geq 0$, $\delta_i \geq 0$.

6.14.   Unconventional Difference Approximations

As long as we are going to utilize finite difference techniques, why not use difference methods which are more closely connected with the original process? Thus, for example, we can consider in place of (6.13.3) or an equivalent equation derived from (6.13.7) the use of the functional equation

$$f(c, T + \Delta) = \min_{v} \ [g(c, v)\Delta + f(c + v\Delta, T)], \tag{6.14.1}$$

where T assumes the values 0, Δ, 2Δ, ..., and $f(c, 0) = 0$. The calculation now proceeds along the lines discussed in Chapter 3. The function $f(c, T)$ can be stored either at grid points in the c-domain or by means of some interpolation function if c is of higher dimension. The use of an interpolation function makes the calculation of $f(c + v\Delta, T)$ for any value of v a relatively simple matter. This latter procedure saves on rapid-access storage, at the expense, as previously pointed out, of computing time.

Observe that there are now no difficulties due to the presence of the ratio of small quantities.

6.15.   Additional Unconventional Difference Approximations

Examination of some of the advantages of the unconventional finite difference scheme in (6.14.1) tempts us to use a similar approach in connection with other classes of partial differential equations. Consider, for example, the equation

$$u_t = uu_x, \qquad u(x, 0) = g(x), \qquad 0 \le x \le 1. \tag{6.15.1}$$

This is a convenient nonlinear partial differential equation on which to test various analytic and computational methods because we possess an analytic representation of the solution of relative simplicity. Specifically, u satisfies the implicit equation

$$u = g(x + tu). \tag{6.15.2}$$

In place of any of the usual finite difference schemes for numerical solution of (6.15.1), consider the equation

$$v(x, t + \Delta) = v(x + v(x, t)\Delta, t),$$
$$v(x, 0) = g(x), \quad 0 \le x \le 1, \tag{6.15.3}$$

where $t = 0$, Δ, 2Δ, ..., and $v(x, t)$ is stored in $0 \le x \le 1$ for each value of t by means of some interpolation formula. The limiting form of (6.15.3) as $\Delta \to 0$ is the partial differential equation in (6.15.1).

Similarly, to treat an equation such as

$$u_t = h(u, x, t)u_x + k(u, x, t), \qquad u(x, 0) = g(x), \qquad (6.15.4)$$

we can use the equation

$$v(x, t + \Delta) = v(x + h(u, x, t)\Delta, t) + k(u, x, t)\Delta, \qquad (6.15.5)$$

$$v(x, 0) = g(x).$$

EXERCISES

1. Show that we can obtain a finite difference approximation accurate to terms of order Δ^3 by means of an appropriate functional equation.
2. Can we obtain approximations agreeing with the original equation to arbitrary powers of Δ?
3. Which might be preferable for numerical purposes: a decrease of Δ with a lower-order formula or the use of a higher-order formula?

6.16. Power Series Expansions

With the development of various computer programs for the algebraic manipulation and differentiation of polynomials, it is feasible to contemplate the local solution of partial differential equations of the form

$$f_T = g(c) - f_c^2/4, \qquad f(c, 0) = 0 \qquad (6.16.1)$$

by means of power series expansions in c or T, or both (where $g(c)$ is a polynomial or rational function of c.)

If we set

$$f = f_1(c)T + f_2(c)T^2 + f_3(c)T^3 + \cdots, \qquad (6.16.2)$$

substitution in (6.16.1) readily yields the relations

$$f_1(c) = g(c),$$

$$f_2(c) = 0,$$

$$f_3(c) = [g(c) - f_1'(c)]^2/3, \qquad (6.16.3)$$

and so forth.

Alternatively, we can consider an expansion in powers of c. Set

$$f = g_1(T)c^2 + g_2(T)c^4 + \cdots,$$

$$g = a_1c^2 + a_2 c^4 + \cdots. \qquad (6.16.4)$$

Substitution in (6.16.1) yields the set of nonlinear ordinary differential equations

$$g_1' = a_1 - g_1{}^2, \qquad g_1(0) = 0,$$
$$g_2' = a_2 - 4g_1 g_2, \qquad g_2(0) = 0, \qquad\qquad (6.16.5)$$
$$\vdots$$

These may be solved recurrently, first g_1, then g_2, and so on, or simultaneously if we decide to calculate $g_1, g_2, \ldots, g_n$. In the multi-dimensional case we may find a sequential approach, plus the linearity of the equations past the first, useful in overcoming dimensionality difficulties.

EXERCISES

1. For the case where $g(c) = c^2 + c^4$, determine the first four functions, $f_i(c)$, $i = 1, 2, 3, 4$, and the equations for the first four functions $g_i(T)$, $i = 1, 2, 3, 4$.
2. Can we readily determine the radius of convergence of the series in (6.16.2)?
3. Can we readily determine the radius of convergence of the series in (6.16.4)?

6.17.   Perturbation Series

We can also utilize the partial differential equation to obtain perturbation expansions in a parameter appearing in the function $g(c)$. If, for example,

$$g(c) = c^2 + \varepsilon g_1(c), \qquad\qquad (6.17.1)$$

we can set

$$f(c, T) = f_0(c, T) + \varepsilon f_1(c, T) + \cdots, \qquad\qquad (6.17.2)$$

and obtain a series of equations for the functions $f_i(c, T)$.

EXERCISES

1. Consider the case where $J(u) = \int_0^T (u'^2 + u^2 + \varepsilon u^4)\, dt$, $u(0) = c$. Compare the series in ε with the series in c. Can we combine the two parameters ε and c into a single parameter?
2. What happens in higher dimensions?

6.18. Existence and Uniqueness

The partial differential equation

$$f_T = \min_v [g(c, v) + vf_c], \qquad f(c, p) = 0, \qquad (6.18.1)$$

is an interesting object in its own right. To investigate it in any depth would take us into the theory of partial differential equations and too far astray from our chosen domain of control theory. Let us, however, make some simple observations.

We have established the existence of a solution of (6.18.1) under appropriate conditions on $g(c, v)$—a solution with the explicit representation

$$f(c, T) = \min_u \int_0^T g(u, u') \, dt, \qquad (6.18.2)$$

where the minimum is taken over functions satisfying the initial condition $u(0) = c$. We shall return to this point below.

There are several ways of studying the question of uniqueness of solution. We shall use an interesting and important result concerning the positivity of an associated linear partial differential operator.

Once an equivalence between the original minimization problem and the solution of the partial differential equation has been established, we possess a number of alternate routes which are independent of the Euler equation.

6.19. Positivity

Consider the linear partial differential equation

$$\frac{\partial v}{\partial T} = g_1(c, T) + h_1(c, T) \frac{\partial v}{\partial c}, \qquad v(c, 0) = 0, \qquad (6.19.1)$$

where $|c| \le c_0$, $0 \le T \le T_0$. Let us assume that g_1 and h_1 satisfy conditions which ensure existence and uniqueness of solution in this region.

It follows in a number of ways, from an explicit solution of (6.19.1), from the use of a directional derivative, or from an appropriate difference approximation (see Section 6.14), that $g_1 \ge 0$ implies $v \ge 0$. A result of this nature is called a *positivity property*. It is more properly characteristic of the inverse of the operator.

If then

$$\frac{\partial w}{\partial T} - h_1(c, T) \frac{\partial w}{\partial c} \ge 0 \qquad (6.19.2)$$

in a region $|c| \le c_0$, $0 \le T \le T_0 \ll 1$, with w_T and w_c such that $w_T - h_1(c, T)w_c$ is an appropriate $g_1(c, T)$, we can conclude that $w \ge 0$ in this region.

6.20.   Uniqueness

Let us use the foregoing positivity result to establish the uniqueness of a solution of

$$f_T = \min_v [g(c, v) + vf_c], \qquad f(c, 0) = 0, \tag{6.20.1}$$

with f_T, f_c continuous in $|c| \leq c_0, 0 \leq T \leq T_0 \ll 1$. Let us suppose that $g(c, v)$ satisfies a condition guaranteeing the existence of the minimum for all $|c| \leq c_0$ and any value of f_c.

Let ϕ be another solution of this equation, satisfying

$$\phi_T = \min_v [g(c, v) + v\phi_c], \qquad \phi(c, 0) = 0. \tag{6.20.2}$$

Furthermore, let $v(c, T)$ be the function yielding the minimum in (6.20.1) and let $w(c, T)$ be the corresponding function for (6.20.2). Then

$$f_T = g(c, v) + vf_c \leq g(c, w) + wf_c, \tag{6.20.3}$$

$$\phi_T = g(c, w) + w\phi_c \leq g(c, v) + v\phi_c.$$

Hence,

$$(f - \phi)_T = f_T - \phi_T \leq w(f_c - \phi_c) = w(f - \phi)_c, \tag{6.20.4}$$

$$\geq v(f_c - \phi_c) = v(f - \phi)_c.$$

It follows from what we have established in Section 6.19 concerning the positivity of the linear partial differential equation that

$$(f - \phi) \leq 0,$$
$$(f - \phi) \geq 0. \tag{6.20.5}$$

Hence,

$$f \equiv \phi. \tag{6.20.6}$$

We see then that the positivity property enables us to establish uniqueness of solution under conditions much weaker than those required for existence.

6.21.   Approximation in Policy Space

The usual approach to the solution of the partial differential equation in (6.20.1) is to use the method of successive approximations after having converted the original equation in some fashion into one of the for

$$f = T(f). \tag{6.21.1}$$

In place of this approach let us employ the powerful method of approximation in policy space. Let $v_0 = v_0(c, T)$ be an initial policy and let f_0 be computed using this policy,

$$\frac{\partial f_0}{\partial T} = g(c, v_0) + v_0 \frac{\partial f_0}{\partial c}, \qquad f_0(c, 0) = 0. \qquad (6.21.2)$$

Let a new policy, v_1, be determined as the function which minimizes the function

$$g(c, v) + v \frac{\partial f_0}{\partial c}. \qquad (6.21.3)$$

Then compute a new function f_1 using this new policy,

$$\frac{\partial f_1}{\partial T} = g(c, v_1) + v_1 \frac{\partial f_1}{\partial c}, \qquad f_1(c, 0) = 0. \qquad (6.21.4)$$

We assert that we obtain monotone approximation in this fashion. Namely,

$$f_1 \leq f_0. \qquad (6.12.5)$$

To see this, observe that the way in which v_1 was obtained implies that

$$\frac{\partial f_0}{\partial T} = g(c, v_0) + v_0 \frac{\partial f_0}{\partial c} \geq g(c, v_1) + v_1 \frac{\partial f_0}{\partial c}. \qquad (6.21.6)$$

Comparing (6.21.2) and (6.21.6), we have the result stated in (6.21.5).

We can continue in this fashion, generating a sequence $\{f_n\}$ with the property that

$$\cdots \leq f_2 \leq f_1 \leq f_0. \qquad (6.21.7)$$

The question of convergence is more complex. What is particularly interesting about the foregoing procedure, however, is that it provides a systematic technique for improving any given approximation.

EXERCISES

1. Apply the method to $f_T = \min_v [c^2 + v^2 + vf_c], f(c, 0) = 0$.
2. Apply the method to $f_T = \min_y [(c, Ac) + (y, y) + (y, \text{grad } f)], f(c, 0) = 0$.

6.22. Quasilinearization

The fact that the functional equations of dynamic programming permit an approach by means of the usual method of successive approximations, as well as by means of approximation in policy space, suggests that it may be worth our while to write an equation of the type

$$u = T(u) \tag{6.22.1}$$

in the form

$$u = \min_v S(u, v), \tag{6.22.2}$$

and take advantage of the fact that v may be regarded as a "policy."

This is the origin of the theory of *quasilinearization*. Let us give an example. Consider the Riccati equation

$$u' = a(t) + u^2 \qquad u(0) = c, \tag{6.22.3}$$

$|c| \ll 1$. Since

$$u^2 = \max_v (2uv - v^2), \tag{6.22.4}$$

we can write this equation in the form

$$u' = \max_v [a(t) + 2uv - v^2]. \tag{6.22.5}$$

This representation of the equation in (6.22.3) permits us to obtain a number of results, as indicated in the exercises at the end of this section. Many further results will be found in the references at the end of this chapter.

EXERCISES

1. Consider the linear differential equation

$$z' = a(t) + 2zv - v^2, \qquad z(0) = c.$$

Write $z = f(v, t)$. Show that $u = \max_v f(v, t)$, both directly and by positivity arguments. Over what t-interval do the results hold?

2. Obtain a corresponding result for $u' = g(u, t)$, $u(0) = c$, where g is uniformly convex in u.

3. How might one obtain *lower* bounds for u?

4. Use the foregoing results to obtain approximate solutions to $u' = a^2 + f(t) + u^2$, $u(0) = c$, where $|f(t)| \ll 1$.

6.23. Representation of Solution of Partial Differential Equation

We have seen that

$$f(c, T) = \min_u \left[\int_0^T g(u, u') \, dt \right], \qquad u(0) = c, \tag{6.23.1}$$

satisfies the equation

$$f_T = \min_v [g(c, v) + vf_c], \qquad f(c, 0) = 0,$$

$$= \phi(c, f_c), \qquad\qquad (6.23.2)$$

under various assumptions. This suggests that we invert our procedure and represent the solution of various equations of the foregoing kind in the form given above.*

Suppose that $\phi(c, w)$ is uniformly concave in w. Then, as we know, we can write

$$\phi(c, f_c) = \min_w [\phi(c, w) + (f_c - w)\phi_w(c, w)]. \qquad (6.23.3)$$

Thus, we have the equation

$$f_T = \phi(c, f_c)$$

$$= \min_w [(\phi - w\phi_w) + \phi_w f_c]. \qquad (6.23.4)$$

Comparing (6.23.2) and (6.23.4), we set

$$v = \phi_w, \qquad\qquad (6.23.5)$$

$$g(c, v) = \phi - w\phi_w.$$

Let us check the convexity of g. We have

$$g(c, v) = \phi - wv,$$

$$g_v = \phi_w w_v - w_v v - w = -w, \qquad (6.23.6)$$

$$g_{vv} = -w_v.$$

From (6.23.5), we have

$$1 = \phi_{ww} w_v. \qquad\qquad (6.23.7)$$

Hence, if $\phi_{ww} < 0$, we have $g_{vv} > 0$.

If the initial condition in (6.23.2) is $f(c, 0) = k(c)$, we replace (6.23.1) by

$$f(c, T) = \min_u \left[k(u(T)) + \int_0^T g(u, u')\, dt \right]. \qquad (6.23.8)$$

An immediate advantage of the representation in (6.23.2) is that any choice of a function u furnishes by way of $\int_0^T g(u, u')\, dt$, $u(0) = c$, upper bounds to the solution of (6.23.4). If we could find a systematic way of obtaining lower bounds for the minimum value in (6.23.1) then we would have a systematic way of obtaining lower bounds for (6.23.4). In Chapter 9, we approach this problem by means of duality.

---

* *Mann immer umkehren muss!*—Jacobi.

6.24.   Constraints

Consider the following variational problem:

$$\min_{u} \int_0^T (u'^2 + u^2)\, dt, \qquad (6.24.1)$$

where

(a) $u(0) = c,$

(b) $|u'| \le k.$ (6.24.2)

Proceeding formally along the lines of Section 6.4, we obtain the equation

$$f_T = \min_{|v| \le k} [c^2 + v^2 + vf_c], \qquad f(c, 0) = 0, \qquad (6.24.3)$$

where, as above,

$$f(c, T) = \min_{u} \left[\int_0^T (u'^2 + u^2)\, dt \right]. \qquad (6.24.4)$$

The question of the existence of solutions of equations of this nature is an interesting and difficult one, concerning which relatively little has been done. Discontinuities of various types occur and there are analogies to shock waves and to the generalized solutions of other types of partial differential equations.

One of the advantages of (6.24.3) is that it can readily be employed to determine the structure of the solution in a heuristic fashion. Once the nature of the solution is known, it can be rigorously derived along different lines.

EXERCISES

1.   Determine the solution of the minimization problem in (6.24.1) for small T.
2.   What happens as T increases?
3.   Consider the minimization of $J(u) = \int_0^T (u'^2 + u^2)\, dt$ subject to $u(0) = c$, $|u'| \le k\,|u|$. Assuming that the minimum is attained, set $f(c, T) = \min_u J(u)$. Show that $f(c, T) = c^2 \phi(T)$. What differential equation does ϕ satisfy, supposing that the analog of (6.24.3) is valid?

6.25.   Inverse Problems

Let us next consider the problem of determining all criterion functionals which produce an observed control law. We shall present the question in the following form: Given the function $v(c, T)$, which determines $f(c, T)$, how do we determine the function, or functions, $g(u, u')$, for which

$$f(c, T) = \min_u \int_0^T g(u, u') \, dt, \qquad u(0) = c? \tag{6.25.1}$$

We can begin with the result already derived that g and v are related by means of the partial differential equation

$$(g - vg_v)_c = -(g_v)_T. \tag{6.25.2}$$

If v is given as a function of c and T, we have a linear second-order partial differential equation for g. Since this presents some difficulties, let us proceed differently. The function f satisfies the equation

$$f_T = g(c, v) + vf_c, \tag{6.25.3}$$

where v and f are related by the equation $g_v + f_c = 0$. Hence,

$$(f_T - vf_c)_T = g_T = g_v v_T = -v_T f_c. \tag{6.25.4}$$

Thus,

$$f_{TT} - v_T f_c - v f_{cT} = -v_T f_c, \tag{6.25.5}$$

whence, finally,

$$f_{TT} - v f_{cT} = 0. \tag{6.25.6}$$

Regard this an an equation for the function f_T,

$$(f_T)_T - v(f_T)_c = 0. \tag{6.25.7}$$

As such, it is a first-order linear equation which can be readily approached using the theory of characterstics

EXERCISES

1. Determine all $g(u, u')$ for which $v(c, T) = ck(T)$.
2. Determine all $g(u, u')$ for which $v(c, T) = \phi(c)k(T)$.
3. Extend the foregoing to the multidimensional case and determine all $g(x, x')$ for which $v(c, T) = A(T)c$.
4. Determine all $g(u, u')$ for which similarity policies exist, i.e., $v(c, T) = \phi(\psi(c)k(T))$.
5. More generally, determine all $g(u, u')$ for whch more general similarity policies exist,

   (a) $v(c, T) = h(c)\phi(\psi(c)k(T))$,

   (b) $v(c, T) = m(T)\phi(\psi(c)k(T))$.

6.26.   Semi-groups and the Calculus of Variations

As was first indicated by Hadamard, a great deal of classical analysis can be interpreted in terms of semi-groups of operators. Many important classes of functional equations can be written in the form

$$u_t = Au, \qquad u(0) = f, \tag{6.26.1}$$

where A is an operator, leading to the symbolic exponential

$$u = e^{At}f. \tag{6.26.2}$$

More generally, we have

$$u_t = Au + g, \qquad u(0) = f. \tag{6.26.3}$$

The dynamic programming approach to the calculus of variations leads to an equation of the form

$$u_t = \min_v (A(v)u + g(v)), \qquad u(0) = f. \tag{6.24.4}$$

Thus the calculus of variations is associated with nonlinear semi-groups as opposed to the linear semi-group of (6.26.1). Equations of the type appearing in (6.26.4) merit investigation on their own.

Miscellaneous Exercises

1. Consider the problem of minimizing $J(u) = \int_0^T g(u, u')\, dt$ subject to $\int_0^T h(u, u')\, dt = k$, where $u(0) = c$. Proceeding formally, write $f(c, k, T) = \min_u J(u)$ and show that $f_T = \min [g(c, v) + vf_c - h(c, v)f_k], f(c, k, 0) = 0$.

2. Consider the same problem using a Lagrange multiplier, $J(u, \lambda) = \int_0^T [g(u, u') + \lambda h(u, u')]\, dt$, $\phi(c, T, \lambda) = \min_u J(u)$. What is the equation for ϕ? What is λ in terms of f? What does the result mean?

3. Consider the minimization of $J(u) = \int_0^T g(u, u')\, dt$ subject to $u(0) = c$, $u(T) = b$. Write $f(c, T) = \min_u J(u)$. What equation and what boundary condition at $T = 0$ does f satisfy.?

4. Consider the minimization of $J(u) = \int_0^T u'^2\, dt$ subject to $\int_0^T g(t)u^2\, dt = 1$, $u(0) = u(T) = 0$, where $g(t) \geq b > 0$. Obtain some associated nonlinear partial differential equations. (*Hints:* (1) Consider $f(c, a) = \min_u \int_a^T u'^2\, dt$, $\int_a^T g(t)u^2\, dt = 1$, $u(a) = c$, $u(T) = 0$, and renormalize as necessary. (2) Consider $f(c, a) = \min_u \int_a^T u'^2\, dt$, $\int_a^T g(t)u^2\, dt = c$, $u(a) = u(T) = 0$, and renormalize as necessary. (See

R. Bellman, *Dynamic Programming of Continuous Processes*, The RAND Corporation, R-271, 1954.

5. Can one obtain corresponding equations for the higher characteristic values?

6. Consider an object, subject only to the force of gravity and air resistance, which is propelled straight up by an initial velocity v. Let the defining equation be $u' = -g - h(u')$, $u(0) = 0$, $u'(0) = v$ and let $f(v)$ denote the maximum altitude attained. Show that

$$f(v) = \int_0^v \frac{v_1 \, dv_1}{g + h(v_1)}.$$

7. Consider the more general case where the motion is through an inhomogeneous medium. Let the defining equation by $u'' = h(u, u')$, $u(0) = c_1$, $u'(0) = c_2$ and let $f(c_1, c_2)$ denote the maximum altitude attained starting with the initial altitude c_1 and initial velocity c_2, $c_2 \geq 0$. Then f satisfies the partial differential equation,

$$c_2 + c_2 \frac{\partial f}{\partial c_1} + h(c_1, c_2) \frac{\partial f}{\partial c_2} = 0, \qquad f(c_1, 0) = 0, \qquad c_1 \geq 0.$$

8. Discuss the use of this relation to determine $f(c_1, c_2)$ numerically and analytically. (See

R. Bellman, "Functional Equations and Maximum Range," *Quart. Appl. Math.*, **17**, 1959, pp. 316–318.)

9. Consider the boundary-value problem $(p(t)u')' + g(t)u = v(t)$, $a < t < 1$, $u(a) = 0$, $u(1) + \alpha u'(1) = 0$, with the solution $u(t) = \int_a^1 k(t, s, \alpha)v(s) \, ds$, and the associated problem of minimizing $J(u, v) = \int_a^1 (q(t)u^2 - p(t)u'^2 - 2uv) \, dt - p(1)u(1)^2/\alpha$. Use the functional equation of dynamic programming to obtain the Riccati-type relation

$$\frac{\partial k}{\partial a}(t, s, a) = p(a) \frac{\partial k}{\partial s}(t, a, a) \frac{\partial k}{\partial t}(a, s, a).$$

10. From this type of relation obtain results for the variation of the characteristic values and functions of the Sturm-Liouville problem $(p(t)u')' + (q(t) + \lambda r(t))u = 0$, $a < t < 1$, $u(a) = 0$, $u(1) + \alpha u'(1) = 0$, with the position of the endpoint a. (*Hint:* Consider $k(t, s, a, \lambda)$ as a meromorphic function of λ and choose appropriate values of λ.)

11. Obtain analogous results for the multidimensional case where the equation is

$$\frac{d}{dt}(P(t)\frac{dx}{dt} + [Q(t)\lambda R(t)]x = y(t), \qquad a < t < 1, \qquad x'(1) + Bx(1) = 0.$$

(See

R. Bellman and R. S. Lehman, "Functional Equations in the Theory of Dynamic Programming—X: Resolvents, Characteristic Functions, and Values," *Duke Math. J.*, **27**, 1960, pp. 55–70.)

12. Consider the minimization of a functional of the form

$$J(u) = \int_0^T g\left(u, u', \int_0^T h(u, u') \, dt_1\right) dt,$$

subject to $u(0) = c$. Introduce the function

$$f(c, a, T) = \min_u \int_0^T \left[g\left(u, u', a + \int_0^T h(u, u')\right) dt\right] dt,$$

where $u(0) = c$, $-\alpha < a$, $c < \infty$, $T \geq 0$. Write $\phi(c, a, T) = \int_0^T h(u, u') \, dt$, where u is the minimizing function. Show formally that

$$f_T = \min_v \left[g(c, v, a + \phi) + vf_c + h(c, v)f_a\right],$$

$$\phi_T = h(c, v) + v\phi_c + h(c, v)\phi_a,$$

with $f(c, a, 0) = 0$, $\phi(c, a, 0) = 0$.

13. Obtain the explicit form of the partial differential equations for the case where

(a) $J(u, a) = \int_0^T \left(a + u' + \int_0^T u \, dt\right)^2 dt + \int_0^T u^2 \, dt,$

(b) $J(u, a) = \int_0^T \left[u'^2 + g\left(a + u + \int_0^T h(u) \, dt\right)\right] dt.$

(See

R. Bellman, "Functional Equations in the Theory of Dynamic Programming XV. Layered Functionals and Partial Differential Equations," *JMAA*, **28**, 1969, pp 1–3.

D. G. B. Edelen, "Problem of Stationarity and Control with a Nonlocal Cost per Unit Time," *JMAA*, **28**, 1969, pp. 660–673.

15. Consider the problem of maximizing $I(y) = \int_0^\infty e^{-at} h(x - y) \, dt$ subject to $x' = g(y)$, $x(0) = c$, $0 \leq y \leq x$, $g(0) = 0$, $g(y) \geq 0$. Write $f(c) = \max_y I(y)$ and show formally that $0 = \max_y \left[h(c - y) - af(c) + g(y)f'(c)\right]$ provided that the maximum occurs inside $[0, c]$.

16. If $g(y) > 0$ for $y > 0$ show that $f(c) = \min_y J(y, g, h)$ where $J(y, g, h)$ is the solution of $0 = (h(c - y)/g(y)) - (af(c)/g(y)) + f'(c)$, $f(0) = 0$, assuming that f satisfies the foregoing nonlinear differential equation.

17. Suppose that we observe that $y = r(c)$, $0 < r(c) < c$ for $c \geq 0$ and the information that this is the optimal policy for a control process of the foregoing nature. How do we determine h, given g? Consider particularly the case where $r(c) = kc$, $0 < k < 1$. (See

   R. Bellman, "Dynamic Programming and Inverse Optimal Problems in Mathematical Economics," *J. Math. Anal. Appl.*, to appear.
   M. Kurz, *On the Inverse Optimal Problem*, Tech. Rep. 3 Institute for Mathematical Studies in the Social Sciences, Stanford University, 1967.)

18. Minimize $\sum_{k=1}^{3} I_k^2 u_k^2 + \lambda \int_0^T (\sum_{i=1}^{3} v_i^2)\, dt$ where $I_1 \dot{u}_1 + (I_3 - I_2)u_2 u_3 = v_1$, $I_2 \dot{u}_2 + (I_1 - I_3)u_3 u_1 = v_2$, $I_3 \dot{u}_3 + (I_2 - I_1)u_1 u_2 = v_3$, $u_i(0) = c_0$, $i = 1, 2, 3$. (See

   T. G. Windeknecht, "Optimal Stabilization of Rigid Body Attitude," *JMAA*, **6**, 1963, pp. 325–335.

19. How would one apply dynamic programming techniques to the solution of the Emden-Fowler-Fermi-Thomas equation $u'' - t^m u^n = 0$, $n > 1$? To the equation $u'' - e^{\lambda t} u^n = 0$? (See

   R. Bellman, "Dynamic Programming and the Variational Solution of the Thomas–Fermi Equation," *J. Phys. Soc., Japan*, **12**, 1957, p. 1049.)

BIBLIOGRAPHY AND COMMENTS

6.1. The problems associated with determining various interconnections between dynamic programming and the calculus of variations have attracted a considerable amount of attention in recent years. See the books

V. G. Boltianskii, *Mathematical Methods of Optimal Control*, Izdat. Nauka, Moscow, 1961; Engl. transl. by Interscience (Wiley), New York, 1962.
C. Carathéodory, *Calculus of Variations and Partial Differential Equations of the First Order —I: Partial Differential Equations of the First Order*, Holden-Day, San Francisco, 1965,
S. E. Dreyfus, *Dynamic Programming and the Calculus of Variations*, Academic Press, New York, 1965.

and

L. D. Berkowitz, "Variational Methods in Problems of Control and Programming," *J. Math. Anal. Appl.*, 3, 1961, pp. 145–169.
V. G. Boltianskii, "Sufficient Conditions for Optimality and the Justification of the Dynamic Programming Method," *J. SIAM Control*, 4, 1966, pp. 326–361.
C. A. Desoer, "Pontriagin's Maximum Principle and the Principle of Optimality," *J. Franklin Inst.*, **27**, 1961, pp. 361–367.
S. E. Dreyfus, "Dynamic Programming and the Hamilton-Jacobi Method of Classical Mechanics," *J. Optimization Theory and Appl.* 2, 1968, pp. 15–26.
A. T. Fuller, "Optimization of Some Nonlinear Control Systems by Means of Bellman's Equation and Dimensional Analysis," *Int. J. Control*, 3, 1966, pp. 359–394.
I. V. Girsanov, "Certain Relations Between the Bellman and Krotov Functions for Dynamic Programming Problems," *J. SIAM Control*, 7, 1969, pp. 64–67.

V. F. Krotov, "Methods of Solving Variational Problems on the Basis of the Sufficient Conditions Governing the Absolute Minimum," *Avtomat. Telemekh.* **23**, 1962; *Automat. and Remote Control*, **23**, 1962, pp. 1473–1484.

P. L. Lukes, "Optimal Regulation of Nonlinear Dynamical Systems," *J. SIAM Control*, **7**, 1969, pp. 75–100.

S. R. McReynolds, "The Successive Sweep Method and Dynamic Programming," *J. Math. Anal. Appl.*, **19**, 1967, pp. 565–598.

R. Morton, "On the Dynamic Programming Approach to Pontriagin's Maximum Principles," *J. Appl. Prob.*, **5**, 1968, pp. 679–692.

V. G. Pavlov and V. P. Cheprasov, "Constructing Certain Invariant Solutions of Bellman's Equation," *Automat. Remote Control*, January, 1968, pp. 31–36.

L. I. Rozonoer, "The Maximum Principle of L. S. Pontryagin in Optimal System Theory," *Automat. Remote Control*, **20**, 1960, pp. 1288–1302; 1517–1532.

A. Tchamram, "A New Derivation of the Maximum Principle," *J. Nath. Anal. Appl.*, **25**, 1969, pp. 350–361.

L. T. Fan et al., "A Sequential Union of the Maximum Principle and Dynamic Programming, *J. Electron. Control*, **XVII**, 1964, pp. 593–600.

For the maximum principle itself, see

L. S. Pontryagin, V. G. Boltyanskii, E. V. Gamkrelidze, and E. F. Mischenko, *Mathematical Theory of Optimal Processes*, Wiley, New York, 1962.

For the derivation of the maximum principle within the framework of the calculus of variations, see

M. Hestenes, *Calculus of Variations and Optional Control Theory*, Wiley, New York, 1966.

An important paper devoted to suboptimal control of nonlinear systems is

A. T. Fuller, "Linear Control of Nonlinear Systems," *Int. J. Control*, **5**, 1967, pp. 197–243.
6.2–5.   See

R. Bellman, *Adaptive Control Processes: A Guided Tour*, Princeton Univ. Press, Princeton, N. J., 1961.

R. Bellman, *Introduction to the Mathematical Theory of Control Processes*, Vol. I: *Linear Equations and Quadratic Criteria*, Academic Press, New York, 1967.
6.7.   The equations derived in this section are equivalent to the classical Hamilton-Jacobi equations. See, for example, the works by Dreyfus cited in Section 6.1. What is quite different, however, and significant is the representation of (6.5.2).
6.8.   See

H. Osborn, "Euler Equations and Characteristics," Chap. 7 in *Dynamic Programming of Continuous Processes*, (R. Bellman, ed.) The RAND Corp. R-271, 1954.
6.9.   See the papers and book by Dreyfus cited in 6.1 and

S. B. Gershwin, "On the Higher Derivatives of Bellman's Equation," *J. Math. Anal. Appl.*, **27**, 1969.

V. W. Merriam, "A Class of Optimum Control Systems, "*J. Franklin Inst.*, **267**, 1959, pp. 267–281.
6.11.   See Volume I.
6.14.   See

R. Bellman and S. Dreyfus, *Applied Dynamic Programming*, Princeton Univ. Press, Princeton, New Jersey, 1962.
6.15.   See

S. Azem, *Higher Order Approximation to the Computational Solution of Partial Differential Equations*, The RAND Corp., RM-3917, 1964.

R. Bellman, "Some Questions Concerning Difference Approximations to Partial Differential Equations," *Boll. UMI*, **17**, 1962, pp. 188–190.

R. Bellman, I. Cherry, and G. M. Wing, "A Note on the Numerical Integration of a Class of Nonlinear Hyperbolic Equations," *Quart. Appl. Math.*, **16**, 1958, pp. 181–183.

R. Bellman, R. Kalaba, and B. Kotkin, "On a New Approach to the Computational Solution of Partial Differential Equations," *Proc. Nat. Acad. Sci. USA.*, **48**, 1962, pp. 1325–1327.

Partial differential equations of this form give rise to some interesting questions. See

M. H. Protter, "Difference Methods and Soft Solutions," *Nonlinear Partial Differential Equations*, Academic Press, New York, 1963, pp. 161–170.

For discussion of the convergence of the solution of the difference equation to the partial differential equation, see

R. Bellman and K. L. Cooke, "Existence and Uniqueness Theorems in Invariant Imbedding —II; Convergence of a New Difference Algorithm," *J. Math. Anal. Appl.*, **12**, 1965, pp. 247–253.

R. Bellman, and R. S. Lehman "Invariant Imbedding, Particle Interaction and Conservation Relations," *J. Math. Anal. Appl.*, **10**, 1965, pp. 112–122.

6.17.   See

R. Bellman, "A Note on Dynamic Programming and Perturbation Theory," in *Nonlinear Vibration Problems*, 1963, pp. 242–244.

6.18.   See

R. Courant and D. Hilbert, *Methods of Mathematical Physics*, Interscience (Wiley, New York, 1953.

6.19.   See Chapter 4 of

E. F. Beckenbach and R. Bellman, *Inequalities*, Springer, Berlin, 1961,

for further discussion and references.

6.21.   For further discussion see the books on dynamic programming cited in Chapter 2, and

J. F. Baldwin and J. H. Sims-Williams, "An On-Line Control Scheme Using a Successive Approximations in Policy Space Approach," *J. Math. Anal. Appl.*, **22**, 1968, pp. 523–536.

6.22.   See

R. Bellman and R. Kalaba, *Quasilinearization and Nonlinear Boundary-value Problems*, Elsevier, New York, 1965.

R. Kalaba, "On Nonlinear Differential Equations, the Maximum Operation, and Monotone Convergence," *J. Math. Mech.*, **8**, 1959, pp. 519–574.

For interesting use of this representation of the solution of the Riccati equations, see

F. Calogero, *Variable Phase Approach to Potential Scattering*, Academic Press, New York, 1967.

See also

R. Bellman, "On Monotone Convergence to Solutions of $u' = g(u, t)$," *Proc. Amer. Math. Soc.*, **8**, 1957, pp. 1007–1009.

6.23.   See the book by Bellman and Kalaba cited in Section 6.22 and

P. D. Lax, "Hyperbolic Systems of Conservation Laws II," *Comm. Pure Appl. Math.*, **10**, 1957, pp. 537–566.

E. D. Conway and E. Hopf, "Hamilton's Theory and Generalized Solutions of the Hamilton-Jacobi Equation," *J. Math. Mech.*, **13**, 1964, pp. 939–986.

   6.24.   See

R. Bellman, I. Glicksbeng and O. Gross, *Some Aspects of the Mathematical Theory of Control Processes*, The RAND Corp., R-313, 1958.

and the works cited in Section 6.1 for discussion of variational problems subject to constraints.

   6.25.   See

R. Bellman and R. Kalaba, "An Inverse Problem in Dynamic Programming and Automatic Control," *J. Math. Anal. Appl.*, **7**, 1963, pp. 322–325,

and references there for the study of these problems within the calculus of variations. See also

R. Bellman, "Dynamic Programming and Inverse Optimal Problems in Mathematical Economics ," *J. Math. Anal. Appl.*, to appear.

and

B. A. Finlayson and L. G. Scriven, "On the Search for Variational Principles," *Int. J. Heat Mass Transfer*, **10**, 1967, pp. 799–821.

   6.26.   For the theory of semi-groups, see

E. Hille and R. Phillips, Functional Analysis and Semigroups, Anal. Math. Soc. Colloq. vol. **XXXI** (1948).

For an extensive discussion of important classes of problems, see

R. E. Beckwith, *Analytic and Computational Aspects of Dynamic Programming Processes of High Dimension*, JPL, California Institute of Technology, Pasadena, 1959, pp. 1–125.

   For a treatment of some implicit problems, see R. Bellman and J. M. Richardson, "On the Application of Dynamic Programming to a Class of Implicit Variational Problems," *Quant. Appl. Math.*, **17**, 1959, pp. 231–236.

7

LIMITING BEHAVIOR OF DISCRETE PROCESSES

7.1. Introduction

In previous chapters we have separately investigated various analytic and computational features of discrete and continuous control processes. In this chapter we wish to examine some interconnections. We shall focus first on the possibility of approximating to a given continuous control process by a discrete control process, a matter of obvious computational significance. Following this, we shall examine the behavior of a class of control processes of discrete type as the time interval between decisions tends to zero.

There are two distinct types of analyses in this latter class, as in the case of descriptive processes. One corresponds to the case where there already exists a continuous process which we can use for purposes of comparison. The other represents a situation where the discrete process is of primary importance and where there need not be a companion continuous process. Both types of investigation are of significance.

In the course of these investigations a number of other interesting questions arise, some of which we shall briefly discuss. Since, however, we are interested only in expounding certain basic ideas and methods, wes hall avoid all temptation to do any extensive exploration of new domains or intensive cultivation of domesticated regions. Many novel types of stability problems arise in the course of these investigations, and we shall allow the reader to judge for himself how much remains to be done.

7.2. Discrete Approximation to the Continuous—and Conversely

The fundamental question we wish to examine is the following. Consider two control processes:

$$x' = g(x, y), \qquad x(0) = c,$$
$$J(x, y) = \int_0^T h(x, y) \, dt, \tag{7.2.1}$$

a continuous process, and

$$x_{n+1} = x_n + g(x_n, y_n)\Delta, \qquad x_0 = c,$$
$$J(\{x_n, y_n\}) = \sum_{n=0}^N h(x_n, y_n)\Delta, \tag{7.2.2}$$

where $N\Delta = T$, a discrete process.

What do we mean when we assert that one is an approximation to the other? More precisely, how many different meanings can we assign to the question? With no loss of generality, we shall consider scalar processes in what follows.

7.3. Suboptimization

One systematic way to proceed is to regard the discrete process as obtained by suboptimization in the continuous process. By "suboptimization" here we mean that we seek the minimum of J over a restricted class of functions or policies rather than over the original set of functions and policies. Unfortunately, we do not immediately obtain in this fashion the specific discrete process of interest, but we come close.

Let us see how this goes. Consider the scalar version of the control process of Section 7.2. and the admissible function $v(t)$, $0 \le t \le T$, defined by

$$v(t) = v_k, \qquad t_k \le t \le t_{k+1}, \qquad t_0 = 0, \qquad t_N = T, \tag{7.3.1}$$

with $u(t)$ defined by the differential equation

$$u' = g(u, v_k), \qquad t_k \le t \le t_{k+1}, \qquad u(0) = c. \tag{7.3.2}$$

Then $J(u, v)$ takes the form

$$J(u, v) = \sum_{k=0}^{N-1} \int_{t_k}^{t_{k+1}} h(u, v_k) \, dt = J(u, \{v_k\}). \tag{7.3.3}$$

By restricting our attention to the determination of the v_k which minimize $J(u, \{v_k\})$ we have obtained a problem which is discrete in the control variable. Clearly,

$$\min_{v} J(u, v) \leq \min_{\{v_n\}} J. \tag{7.3.4}$$

7.4. Lower Bound

On the other hand, if a minimizing pair u and v exist, we have, using these functions and the mean-value theorem,

$$\int_0^T h(u, v)\, dt = \sum_{k=0}^{N-1} \int_{t_k}^{t_{k+1}} h(u, v)\, dt = \sum_{k=0}^{N-1} \int_{t_k}^{t_{k+1}} h(u, v_k + (t - t_k)\theta_k)\, dt, \tag{7.4.1}$$

with

$$u' = g(u, v_k + (t - t_k)\theta_k), \qquad t_k \leq t \leq t_{k+1}, \qquad u_0 = c. \tag{7.4.2}$$

Expanding g and h, by using the bounds for u and v, we further obtain

$$u' = g(u, v_k) + 0(t_{k+1} - t_k),$$

$$\int_0^T h(u, v)\, dt = \sum_{k=0}^{N-1} \left[\int_{t_k}^{t_{k+1}} h(u, v_k)\, dt + 0(t_{k+1} - t_k)^2 \right], \tag{7.4.3}$$

where the 0-terms are uniform.

If we take $|t_{k+1} - t_k| \leq \Delta$, $N\Delta = T$, we see that we have

$$J(u, v) = J(\{v_n\}) + 0(\Delta), \tag{7.4.4}$$

for a specific set of $\{v_n\}$, with

$$u' = g(u, v_k) + 0(\Delta), \qquad t_k \leq t \leq t_{k+1}, \qquad u_0 = c. \tag{7.4.5}$$

Using the type of stability analysis we shall discuss in the next chapter, under reasonable assumptions concerning g and h, we can demonstrate that it is permissible to neglect the uniform $0(\Delta)$ term in (7.4.5) and still maintain (7.4.4). We see then that

$$J(u, v) \geq J(\{v_n\}) - 0(\Delta) \geq \min_{\{v_n\}} J(\{v_n\}) - 0(\Delta). \tag{7.4.6}$$

Combining this with (7.3.4), we have the desired result,

$$\min_{\{v_n\}} J(\{v_n\}) \geq \min_{v} J(u, v) \geq \min_{\{v_n\}} J(\{v_n\}) - 0(\Delta). \tag{7.4.7}$$

We have made strong use of the knowledge that a solution of the continuous variational problem exists.

7.5. Further Reduction

If we now suppose that we possess some a priori bounds on the v_k, either imposed or derived, we can obtain a further simplification. We write, using (7.4.3) and the mean-value theorem once again,

$$\int_{t_k}^{t_{k+1}} h(u, v_k + (t - t_k)\theta_k)\, dt = h(u_k, v_k)(t_{k+1} - t_k) + 0(t_{k+1} - t_k)^2, \quad (7.5.1)$$

and using (7.4.2),

$$u_{k+1} = u_k + g(u_k, v_k)\,(t_{k+1} - t_k). \tag{7.5.2}$$

If we choose $t_k = k\Delta$, we come quite close to the form in (7.2.2).

Again a rudimentary stability analysis is required to yield the desired approximation between the two processes in Section 7.2. We do not go into details since these results play no subsequent role in what follows, and they are, again, results of local nature in time and function space.

7.6. Linear Equations and Quadratic Criteria

The results are naturally most complete and satisfying when the describing equation is linear and the criterion is quadratic. In this case we can readily obtain explicit analytic representations of the minimizing functions and the corresponding criterion functions, thereby permitting us to verify that, by direct examination of the quantities involved, we have the desired approximations. The analytic basis for this approach may be found in Volume I.

EXERCISE

1. Can we obtain results of the foregoing type for more general processes by using successive approximations involving linear equations and quadratic criteria in both the discrete and continuous control processes?

7.7. Sophisticated Quadrature

In the foregoing pages we obtain a discrete control process by using a simple quadrature formula,

$$\int_0^T f(t)\, dt \cong \sum_{k=0}^{T-1} f(k\Delta)\Delta. \tag{7.7.1}$$

We could obtain a far more accurate approximation by using the trapezoidal formula. Or we could employ a still more sophisticated quadrature formula, say that due to Gauss,

$$\int_0^T f(t)\, dt \cong \sum_{k=0}^{N-1} w_k f(t_k), \tag{7.7.2}$$

where the t_k and w_k are chosen so that the two expressions are equal for polynomials of degree $2N - 1$ or less.

The t_k are now not equally spaced. It is with an approximation of this type in mind that we carried out the analysis, in terms of a general set of points $\{t_i\}$, in Sections 7.3–7.5. Since the w_k, the Christoffel numbers, vary with k, the discrete control process obtained in this fashion is one where the return per stage is time-dependent. Thus, we see that we can obtain a better approximation in some sense to a stationary continous control process by means of a nonstationary discrete control process than by a stationary discrete control process. There is clearly a strong motivation to keep the number of quadrature points as small as possible.

7.8. Degree of Approximation

In our discussion in Sections 7.3–7.5 we measured the degree of approximation of one process to another in terms of the closeness of the values of the return functions. What can we conclude about the minimizing functions? Sometimes this is an important question and sometimes not. In a genuine control process as long as we can come satisfactorily close to the actual minimum cost, we may not be very concerned about the relation between the the actual optimal policy or minimizing function and the approximate policy or function. In many cases, economic or engineering considerations dictate the type of suboptimal policy that is desirable, which is to say the class of approximations we may use.

If, however, we are employing the Rayleigh-Ritz artifice to treat a particular equation, an equation arising from a descriptive process, then the question is basic. Suppose, for example, that we are using the minimization of

$$J(u) = \int_0^T [u'^2 + 2g(u)]\, dt \tag{7.8.1}$$

as a tool to study the solution of

$$u'' - g'(u) = 0, \qquad u(0) = c, \qquad u'(T) = 0. \tag{7.8.2}$$

We shall assume in what follows that either g is convex or else quadratic with the required positive definite character, so that (7.8.2) has a unique solution, which furnishes the minimum value in (7.8.1).

Let v be a function for which

$$J(v) - J(u) \leq \varepsilon. \tag{7.8.3}$$

We want to estimate $|v - u|$ in terms of ε.

7.9. Quadratic Case

As long as we are considering the quadratic case, we may just as well allow time-dependence. Let

$$J(u) = \int_0^T (u'^2 + a(t)u^2) \, dt \tag{7.9.1}$$

and impose the condition that $a(t)$ is such that J is positive definite for functions satisfying the conditions

$$u(0) = 0, \qquad u'(T) = 0. \tag{7.9.2}$$

As we know from Volume I, Chapter 4, the foregoing ensures the existence of a unique solution of

$$u'' - a(t)u = 0, \qquad u(0) = c, \qquad u'(T) = 0. \tag{7.9.3}$$

Let v be a function, satisfying the foregoing boundary conditions, such that

$$J(v) - J(u) \leq \varepsilon. \tag{7.9.4}$$

It follows that

$$J(v) = J(u + v - u) = J(u) + J(v - u).* \tag{7.9.5}$$

Hence, (7.9.4) implies that

$$J(v - u) \leq \varepsilon. \tag{7.9.6}$$

We now invoke the positive definiteness of J for a function of the form $v - u$. By assumption, there exists a constant k_3 such that

$$J(v - u) = \int_0^T [(v' - u')^2 + a(t)(v - u)^2] \, dt \geq k_3 \int_0^T [(v' - u')^2 + (v - u)^2] \, dt. \tag{7.9.7}$$

The Cauchy-Schwarz inequality yields

$$\int_0^T [(v' - u')^2 + (v - u)^2] \, dt \geq \int_0^t [(v' - u')^2 + (v - u)^2] \, dt$$

$$\geq \left| \int_0^t (v' - u')(v - u) \, dt_1 \right|. \tag{7.9.8}$$

---

* Recall the middle terms vanish as a consequence of an integration by parts.

Thus,

$$\varepsilon \geq k_3(v - u)^2.$$

Whether the bound is useful depends upon the constant k_3.

EXERCISES

1. Consider the same question for $J(u) = \int_0^T [u'^2 + a(t)u^2 + 2b(t)u] \, dt$.
2. Let u be the solution of $u'' - a(t)u = 0$, $u(0) = c$, $u'(T) = 0$. Show that $u'(0) = -\int_0^T (u'^2 + a(t)u^2) \, dt/c$.
3. What is a corresponding result in the multidimensional case?
4. Carry through the analysis of the foregoing section for the multidimensional case. How can one use this result to obtain an approximate solution of the original equation?

7.10. Convex Case

Let us next consider the case where the function $g(u)$ appearing in (7.8.1) is a convex function. We have

$$J(v) - J(u) = \int_0^T [v'^2 - u'^2 + 2g(v) - 2g(u)] \, dt_1. \tag{7.10.1}$$

Once again, write $v = u + (v - u)$, and suppose that $u(0) = v(0)$. Then

$$v'^2 = u'^2 + 2u'(v' - u') + (v' - u')^2,$$

$$g(v) = g(u) + (v - u)g'(u) + (v - u)^2 g''(\theta)/2, \tag{7.10.2}$$

using the mean-value theorem.

We thus obtain

$$J(v) - J(u) = \int_0^T [(v' - u')^2 + g''(\theta)(v - u)^2] \, dt_1, \tag{7.10.3}$$

since the terms which are linear in $(v' - u')$ and $(v - u)$ drop out upon an integration by parts. If we use the fact that $g''(\theta) \geq 0$ (by hypothesis), we have

$$\varepsilon \geq J(v) - J(u) = \int_0^T [\cdots] \, dt_1 \geq \int_0^t (v' - u')^2 \, dt. \tag{7.10.4}$$

The Cauchy-Schwarz inequality yields

$$|v - u|^2 = \left| \int_0^t (v' - u') \, dt_1 \right|^2 \leq t \int_0^t (v' - u')^2 \, dt_1 \leq T\varepsilon. \tag{7.10.5}$$

EXERCISES

1. If we suppose that $g''(u) \geq k_1 > 0$ for all u, what upper bound do we obtain for $|v - u|$? Is this always an improvement over (7.10.5)?

2. How would we approach the problem of estimation in the case where $J(u) = \int_0^T g(u, u') \, dt_1$?

3. Can we avoid the use of the mean-value theorem? Is there any advantage to the use of the mean-value theorem?

7.11.   Deferred Passage to the Limit

We have shown in the foregoing sections, or at least indicated how one would go about demonstrating, that under reasonable assumptions of smoothness, the determination of the minimum of a functional such as

$$J(u) = \int_0^T g(u, u') \, dt \tag{7.11.1}$$

can be estimated accurately in terms of the minimum of the function

$$J(\{u_n\}) = \sum_{n=0}^{N} g(u_n, v_n)\Delta, \tag{7.11.2}$$

where

$$u_{n+1} = u_n + v_n \Delta, \qquad u_0 = c, \tag{7.11.3}$$

and $\Delta \ll 1$. Writing

$$f(c, T) = \min_{u} J(u),$$

$$\tag{7.11.4}$$

$$f_N(c) = \min_{\{v_n\}} J(\{v_n\}),$$

we derived, under suitable assumption, an inequality of the form

$$|f(c, T) - f_N(c)| \leq k_1\Delta. \tag{7.11.5}$$

Suppose that we want a more accurate estimate of $f(c, T)$ than that provided by $f_N(c)$ for a specific value of Δ. A direct remedy, of course, is to decrease the value of Δ. The difficulty faced by this immediate response, however, is that the time required to calculate $f_N(c)$ is essentially directly proportional to $1/\Delta$, assuming that we are using a dynamic programming algorithm to calculate $f_N(c)$. If $\Delta = 0.01$ and if we decide to increase the accuracy by replacing Δ by Δ^2, we face the fact that the new calculation will take about one-hundred times as long as the original calculation. This is not a very satisfactory state of affairs.

With $g(u, u')$ sufficiently differentiable we can do much better. In place of an estimate such as (7.11.5), we can derive an expansion

$$f_N(c) = f(c, T) + f_1(c, T)\Delta + f_2(c, T)\Delta^2 + \cdots, \qquad (7.11.6)$$

for small Δ by means of an extension of the foregoing error analysis. Let us suppose that this has been done. Write

$$f_N(c) \equiv f_N(c, \Delta), \qquad (7.11.7)$$

and consider the effect of replacing Δ by $\Delta/2$. The result is

$$f_N(c, \Delta/2) = f(c, T) + f_1(c, T)\Delta/2 + f_2(c, T)\Delta^2/4 + \cdots. \qquad (7.11.8)$$

It follows that

$$2f_N(c, \Delta/2) - f_N(c, \Delta) = f(c, T) + f_2(c, T)\Delta^2/2 + \cdots. \qquad (7.11.9)$$

Thus, the expression $2f_N(c, \Delta/2) - f_N(c, \Delta)$ furnishes an estimate for $f(c, T)$ which is accurate to $0(\Delta^2)$. The effort required to evaluate this expression is approximately three times as long as that required to evaluate $f_N(c, \Delta)$, i.e., directly proportional to $1/\Delta$ and $2/\Delta$. We therefore can obtain much greater accuracy at a very small increase in computing time, provided the function g is well enough behaved.

EXERCISES

1. Establish the result in (7.11.6) for the case where $J(u) = \int_0^T (u'^2 + u^2)\, dt$.
2. Establish the result in (7.11.6) for the case where $u' = au + v$, $J(u, v) = \int_0^T (u^2 + v^2)\, dt$.
3. Can we obtain rigorous results of this nature based upon the joint use of successive approximations involving linear equations for the relationship between u, u and v, and quadratic expressions for the functionals?

7.12.   Use of Analytic Structure

The basic idea guiding our analysis is that the greater the knowledge we have of the analytic structure of the solution, the easier it is to obtain accurate numerical results. The use of nonlinear extrapolation in Chapter 3 was another instance of this fundamental concept.

As we repeatedly emphasized, the effective utilization of analog and digital computers requires both analysis and ingenuity. It accentuates the need for sophisticated mathematics.

7.13. Self-consistent Convergence

Let us now turn to the problem of ascertaining the limiting behavior of a discrete process when the time interval between decisions tends to zero. This is a stability question, which involves the structure of the discrete control process.

In the previous sections we started from the springboard of a continuous process. Here we are interested in a self-consistent approach. We start with a discrete control process and seek conditions which assure the convergence of the return function as $\Delta \to 0$ where Δ is the time interval between decisions. What is rather interesting is that convergence holds under very weak assumptions concerning $g(u, v)$—the single-stage return. These assumptions are so mild that they are not sufficient to guarantee the existence of an associated continuous process.

We have not examined the deeper problems connected with the convergence of policies. These questions require both more stringent assumptions and more detailed analysis.

7.14. Precise Formulation

Consider the discrete control process defined by

$$u_{k+1} = u_k + v_k \Delta, \qquad u_0 = c,$$

$$J(\{v_k\}) = \sum_{k=0}^{N} g(u_k, v_k)\Delta, \tag{7.14.1}$$

where the v_k are subject to the constraint

$$|v_k| \leq m, \qquad k = 0, 1, \ldots, N - 1, N. \tag{7.14.2}$$

We take $N\Delta = T$, and we wish to examine the convergence of the minimum of J as Δ approaches zero with T fixed. This means that the number of stages, N, becomes unbounded.

We shall consider only the foregoing simple case to illustrate the approach, leaving the case of time-dependence, terminal control, and more general describing equations such as

$$u_{k+1} = u_k + h(u_k, v_k)\Delta \tag{7.14.3}$$

as exercises of lesser or greater degrees of difficulty.

7.15. Lipschitz Conditions

In order to establish the desired convergence, we require a Lipschitz condition on $g(c, v)$ as a function of c uniform over the allowed v-region. Let us suppose that

$$|g(c, v) - g(c_1, v)| \leq m_2 |c - c_1|^a, \tag{7.15.1}$$

for some $a > 0$, in the region $-m_1 \leq c, c_1 \leq m_1, |v| \leq m$.

Let us consider a subinterval, $|c| \leq m_3$, where m_3 is chosen so that $m_3 + N\Delta \leq m_1$. This means that if c is restricted to this subinterval when there are N stages remaining, at each of the remaining stages the state variable will stay in the region where (7.15.1) holds. We could, if we wish, impose a different constraint on v—one that forces the state variable to remain in the fixed interval $|c| \leq m_1$.

Set

$$f_n(c) = \min_{\{v_k\}} \left[\sum_{k=0}^n g(u_k, v_k)\Delta \right], \qquad n = 0, 1, \ldots, N, \tag{7.15.2}$$

so that

$$f_0(c) = \min_v [g(c, v)\Delta],$$
$$f_{n+1}(c) = \min_v g[(c, v)\Delta + f_n(c + v\Delta)], \qquad n \geq 0. \tag{7.15.3}$$

We wish to show that (7.15.1) yields a uniform Lipschitz condition

$$|f_n(c) - f_n(c_1)| \leq Tm_2 |c - c_1|^a, \qquad n = 0, 1, \ldots, \tag{7.15.4}$$

for c and c_1 in the admissible interval. The proof is inductive. Suppose that we have established

$$|f_n(c) - f_n(c_1)| \leq k_n |c - c_1|^a. \tag{7.15.5}$$

For $n = 0$, we have $k_0 = 1$. Then, proceeding as in Chapter 2

$$f_n(c) = g(c, v_1) + f_{n-1}(c + v_1\Delta) \leq g(c, v_2) + f_{n-1}(c + v_2\Delta),$$
$$f_n(c_1) = g(c_1, v_2) + f_{n-1}(c_1 + v_2\Delta) \leq g(c_1, v_1) + f_{n-1}(c_1 + v_1\Delta), \tag{7.15.6}$$

where v_1 and v_2 are, respectively, the minimizing values in (7.15.3) for c and c_1.

From (7.15.6) follows

$$f_n(c) - f_n(c_1) \leq [g(c, v_1) - g(c_1, v_1)]\Delta + f_{n-1}(c + v_1\Delta) - f_{n-1}(c_1 + v_1\Delta)$$
$$\geq [g(c, v_2) - g(c_1, v_2)]\Delta + f_{n-1}(c + v_2\Delta) - f_{n-1}(c_1 + v_2\Delta), \tag{7.15.7}$$

whence

$$|f_n(c) - f_n(c_1)| \leq \max \, [|g(c, v_1) - g(c_1, v_1)|\Delta$$
$$+ |f_{n-1}(c + v_1\Delta) - f_{n-1}(c_1 + v_1\Delta)|, \, |g(c, v_2) - g(c_1, c_2)|\Delta$$
$$+ |f_{n-1}(c + v_2\Delta) - f_{n-1}(c + v_2\Delta)|]. \qquad (7.15.8)$$

Using (7.15.1) and the inductive hypothesis in (7.15.5), we have

$$|f_n(c) - f_n(c_1)| \leq m_2 \, |c - c_1|^a \Delta + k_{n-1} \, |c - c_1|^a. \qquad (7.15.9)$$

We see that we may take

$$k_n = m_2 \, \Delta + k_{n-1}, \qquad (7.15.10)$$

whence $k_n = nm_2 \, \Delta \leq N\Delta m_2 = Tm_2$. Thus, (7.15.4) is established.

EXERCISE

1. Why do we not bother to say $a \leq 1$ in the Lipschitz condition of (7.15.1)?

7.16. An Intermediate Process

Let us next consider the control process obtained by replacing Δ by 2Δ. To this end, we should write

$$f_n(c) = f_n(c, \Delta), \qquad (7.16.1)$$

to denote the fact that $f_n(c)$ is calculated using the time step Δ. For the moment, however, to simplify the notation a bit, let us set

$$g_n(c) \equiv f_n(c, 2\Delta). \qquad (7.16.2)$$

Eventually, we wish to compare $f_n(c, 2\Delta)$ with $f_{2n}(c, \Delta)$. It is intuitively clear that if $\Delta \ll 1$, these two functions should be close, and our proof will be based upon this comparison.

To simplify this comparison, we construct an intermediate control process with time step Δ. The original process of Section 7.14 is modified by allowing only suboptimal policies in which the same choice of v is made two stages at a time. Thus, we introduce the sequence $\{h_{2n}(c)\}$ defined by

$$h_0(c) = \min_v \, [g(c, v)\Delta + g(c + v\Delta, v)\Delta],$$

$$h_{2n}(c) = \min_v \, [g(c, v)\Delta + g(c + v\Delta, v)\Delta + h_{2n-2}(c + 2v\Delta)], \quad (7.16.3)$$

$n \geq 0$. It is clear that

$$f_{2n}(c) \leq h_{2n}(c). \qquad (7.16.4)$$

It remains to compare $g_n(c)$ and $h_{2n}(c)$.

7.17. Comparison of g_n and h_{2n}

We have

$$g_n(c) = \min_v [2g(c, v)\Delta + g_{n-1}(c + 2v\Delta)]. \tag{7.17.1}$$

Comparing this with (7.16.3), we obtain as before

$$|g_n(c) - h_{2n}(c)| \leq \max_{c,v} |g(c, v) + g(c + v\Delta) - 2g(c, 2v\Delta)|$$

$$+ \max_c |g_{n-1}(c) - h_{2n-2}(c)|, \tag{7.17.2}$$

Hence, we can conclude as above that

$$|g_n(c) - h_{2n}(c)| \leq Tm_4 \Delta^a. \tag{7.17.3}$$

7.18. $f_N(c, \Delta)$ as a Function of Δ

For each value of Δ let N be determined by $N\Delta = T$ and let Δ be replaced by $\Delta/2$, $\Delta/4$, and so forth. Write

$$u_r(c) = f_N(c, \Delta/2^r). \tag{7.18.1}$$

We see that (7.16.4) and (7.17.3) yield

$$f_{2n}(c) \leq h_{2n}(c) \leq g_n(c) + Tm_4 \Delta^a. \tag{7.18.2}$$

Thus, setting $n = N/2$,

$$u_2(c) \leq u_1(c) + Tm_4 \Delta^a. \tag{7.18.3}$$

Generally, carrying through the same analysis for $\Delta/2^r$, we obtain

$$u_{r+1}(c) \leq u_r(c) + Tm_4 \Delta^a 2^{-ra}. \tag{7.18.4}$$

This is almost monotone convergence; indeed it is sufficient to ensure convergence as the next section shows.

7.19. Almost Monotone Convergence

The convergence of $\{u_r(c)\}$ follows from the following lemma:

LEMMA. *If $\{a_n\}$ is a bounded sequence such that*

$$a_{n+1} \leq a_n + b_n, \qquad n = 0, 1, \ldots, \tag{7.19.1}$$

where $\sum b_n$ converges, then a_n approaches a limit as $n \to \infty$.

The proof is simple. Set

$$a_n = c_n + \sum_{k=0}^{n-1} b_k.$$ (7.19.2)

Then (7.19.1) becomes

$$c_{n+1} \le c_n.$$ (7.19.3)

Since $\sum_k b_k$ is convergent, the boundedness of a_n ensures the boundedness of c_n. The usual result on monotone convergence can now be applied.

7.20. Discussion

The importance of the foregoing method is that it can readily be applied to a number of discrete processes of complex type where it is quite difficult to formulate an associated continuous process. We are thinking in particular of stochastic and adaptive processes.

BIBLIOGRAPHY AND COMMENTS

7.1. Problems of the type treated in this chapter are part of stability theory. See

R. Bellman, *Methods of Nonlinear Analysis*, Academic Press, New York, 1970.

R. E. Beckwith, *Analytic and Computational Aspects of Dynamic Programming Processes of High Dimension*, JPL, California Institute of Technology, 1959.

B. M. Budak, E. M. Berkovich and E. N. Soloveva, "Difference Approximations in Optimal Control Problems," *J. SIAM Control*, 7, 1969, pp. 18–31.

J. Cullum, "Discrete Approximations to Continuous Optimal Control," *J. SIAM Control*, 7, 1969 pp. 37–49.

W. H. Fleming, *A Resource Allocation Problem in Continuous Form*, The RAND Corp., RM-1430, 1955.

W. H. Fleming, *Discrete Approximations to Some Differential Games*, The RAND Corp., RM-1526, 1955.

7.2. The advantage of suboptimization is that one obtains an immediate bound, upper or lower as the case may be.

7.7. A detailed discussion of quadrature will be given subsequently in Chapter 12.

7.8–10. See the Bellman book cited in 7.1.

7.11. See, for example,

V. Pereyra, "On Improving an Approximate Solution of a Functional Equation by Deferred Corrections," *Numer. Mat.* 8, 1966, pp. 376–391.

7.13. Results of this nature were first presented in

R. Bellman, "Functional Equations in the Theory of Dynamic Programming—VI: A Direct Convergence Proof," *Ann. Math.*, 65, 1957, pp. 215–223.

8

———

ASYMPTOTIC CONTROL THEORY

8.1. Introduction

One of the principal objectives of the theory of differential equations is to provide a systematic approach to the discovery of properties of solutions of equations of the form

$$x' = g(x), \qquad x(0) = c. \tag{8.1.1}$$

This initial-value problem has a unique solution in some interval $[0, t_0]$ provided that $g(x)$ satisfies a Lipschitz condition in the neighborhood of c. The problem, however, of ascertaining when x exists for all $t \geq 0$ and if it does, the asymptotic behavior of x as $t \to \infty$ is a very difficult one. A reasonably comprehensive theory exists in the case where $g(x)$ is of the form

$$g(x) = Ax + h(x), \tag{8.1.2}$$

where $\|h(x)\|$ is small when $\|x\|$ is small; more precisely, $\|h(x)\| = o(\|x\|)$ as $\|x\| \to 0$, A is a stability matrix,* and $\|c\| \ll 1$.

The fundamental result of Poincaré and Lyapunov is that under these assumptions a great deal of the asymptotic behavior of x can be predicted on the basis of the asymptotic behavior of the solution of the associated linear equation

$$\frac{dy}{dt} = Ay, \qquad y(0) = c. \tag{8.1.3}$$

---

* By this we mean that all of the characteristic roots of A have negative real parts.

We would like to obtain similar results for control processes and for two-point boundary-value problems in general. Although results of this nature are more difficult to obtain if only because two-point boundary-value problems automatically introduce a multitude of complexities, a certain amount of progress can be made. In the first part of this chapter we shall present some theorems which can be considered to be analogues of the classical ones for initial-value equations.

At the conclusion of this chapter we shall consider asymptotic behavior of a quite different nature for solutions of equations of the form

$$f_n(p) = \max_q \, [g(p, q) + f_{n-1}(T(p, q))]. \tag{8.1.4}$$

The method employed has some independent interest.

8.2. Asymptotic Control

Let us now describe some kinds of problems arising in the area of what we may call "asymptotic control." Consider the familiar question of determining the minimum of the functional

$$J(u) = \int_0^T (u'^2 + 2g(u)) \, dt, \tag{8.2.1}$$

subject to $u(0) = c$ and the condition that $u' \in L^2(0, T)$. Let us suppose that g is convex. This ensures that the minimum value exists and that the minimizing function is furnished by the unique solution of

$$u'' - g'(u) = 0, \qquad u(0) = c, \qquad u'(T) = 0. \tag{8.2.2}$$

The function $u = u(t, T)$ exists for all $T > 0$, $0 \le t \le T$. Consequently, it is of some interest to ask about the behavior of the minimum value and minimizing function as $T \to \infty$.

Write, as usual,

$$f(c, T) = \min_u J(u), \tag{8.2.3}$$

and $u(t, T)$ for the solution of (8.2.2). There are then a number of questions that we can ask:

(a)   Does $f(c, T)$ possess a limit as $T \to \infty$?
(b)   Does $u(t, T)$ possess a limit as $T \to \infty$?
(c)   Assuming that these limit functions exist, what connection is there, if any, between these functions and the problem of minimizing the functional $\int_0^\infty (u'^2 + 2g(u)) \, dt$.

(d)   What connection is there with the solution of $u'' - g'(u) = 0$, $u(0) = c$, $\lim_{t \to \infty} u = 0$?

(e)   In general, can we determine the asymptotic behavior of $u(t, T)$ and $f(c, T)$ as $T \to \infty$?

(f)   Can we use the solution over the infinite interval as the first term in the asymptotic expansion of the solution of the finite problem with large T?

The last question reveals a principal motivation in the study of asymptotic control theory. Our aim is to use the "steady-state" control associated with an infinite process as an approximation to the optimal policy for a finite process of long duration.

We shall touch on only a few of these matters since any detailed discussion would consume a considerable amount of time and space.

8.3.   Existence of Limit of $f(c, T)$

Let us begin by demonstrating the existence of the limit of $f(c, T)$ as $T \to \infty$ under certain general conditions. We suppose that

(a) $g \geq 0$,

(b) $\displaystyle\int_0^\infty g(ce^{-t})\, dt < \infty$

$$(8.3.1)$$

for any c, and we continue the previous assumption that g is convex.

It is clear, first of all, that the assumptions in (8.3.1) imply that $f(c, T)$ is monotone increasing in T. Furthermore, using the trial function $u = ce^{-t}$, we have

$$f(c, T) \leq \int_0^T [c^2 e^{-2t} + 2g(ce^{-t})]\, dt$$

$$\leq \int_0^\infty [\cdots]\, dt < \infty,$$

$$(8.3.2)$$

by virtue of assumption in (8.3.1b). It follows that $f(c, T)$ converges as $T \to \infty$.

EXERCISES

1.   Consider the cases where $J(u) = \int_0^T (u'^2 + a^2 u^2)\, dt$. Show by direct examination of $f(c, T)$ and $u(t, T)$ that these functions possess limits as $T \to \infty$.

2.   Obtain estimates of $f(c, T) - f(c, \infty)$ and $u(t, T) - u(t, \infty)$.

3.   Establish similar results for $J(x) = \int_0^T [(x', x') + (x, Ax)]\, dt$, where $A > 0$.

4.   Show that $(c, A^{1/2}c) = \lim_{T \to \infty} \min_x J(x)$, where $A^{1/2}$ is the positive definite square root of A.

5.   Hence, show that $(A + B)^{1/2} \geq A^{1/2} + B^{1/2}$ if A, $B > 0$.

6. What meanings can we assign to the minimization of

$$J_1(u) = \int_0^\infty (u'^2 + a^2 u^2)\, dt, \quad u(0) = c,$$

and what is the relation with $\lim_{T\to\infty} J(u)$?

8.4. Poincaré-Lyapunov Theory

In order to motivate the discussion that follows concerning two-point boundary-value problems, let us recall the classical result of Poincaré and Lyapunov in stability theory. Consider the initial-value problem

$$x' = Ax + g(x), \qquad x(0) = c, \tag{8.4.1}$$

where we assume the following:

(a) A is a stability matrix,
(b) $\|g(x)\| = o(\|x\|)$ as $\|x\| \to 0$, $\tag{8.4.2}$
(c) $\|c\| \ll 1$.

Then under these conditions we can assert that any solution of (8.4.1) may be continued indefinitely and approaches zero as $t \to \infty$. If we add a Lipschitz condition,

$$\|g(x) - g(y)\| \le k_1 \|x - y\| \tag{8.4.3}$$

for $\|x\|, \|y\| \ll 1$, we can assert that *the* solution of (8.4.1) can be continued indefinitely in t, and that $x \to 0$ as $t \to \infty$. Furthermore, we can show that $x \sim \exp(\lambda_1 t)\, b$ as $t \to \infty$ where λ_1 is the characteristic root of A with largest real part, assuming that this is a simple root where b is determined by c. A corresponding result can be obtained for the case where λ_1 is a multiple root.

Another way of saying the foregoing is that the solution of (8.4.1) behaves asymptotically like a solution of the linear approximation

$$y' = Ay, \tag{8.4.4}$$

as $t \to \infty$.

One proof of this basic result hinges upon the conversion of (8.4.1) into the nonlinear integral equation

$$x = e^{At}c + \int_0^t e^{A(t-t_1)}g(x)\, dt_1. \tag{8.4.5}$$

Application of the method of successive approximations along the lines indicated in the following exercises yields the stated result. The proof is only sketched here since we shall go through the corresponding proof in detail in the next section.

EXERCISES

1. Set $x_{n+1} = \exp(AT) c + \int_0^t \exp(A(t - t_1)) g(x_n) dt_1$, $n \geq 0$, with $x_0 = c$. Show that the hypotheses enable us to assert that $\|x_n\| \leq 2\|c\|$.

2. Show that $\sum_n \|x_{n+1} - x_n\|$ converges, and thus that $x_n \to x$, a solution of (8.4.5), as $n \to \infty$.

3. Show that $x \sim \exp(\lambda_1 t) b$ as $t \to \infty$. (*Hint*: Prove first inductively that $\|x\| \leq b_1 \exp(\text{Re}(\lambda_1) t)$ and then write $\int_0^t = \int_0^\infty - \int_t^\infty$.)

8.5. Analogous Result for Two-point Boundary-value Problem

We wish to obtain a corresponding result for some classes of two-point boundary-value problems. It is sufficient to consider the scalar equation

$$u'' - u = h(u), \qquad u(0) = b. \qquad u'(T) = 0, \qquad (8.5.1)$$

to illustrate the general method. We use the initial vector b instead of c to avoid some notational conflict. To begin with, we ignore possible connections with the functional

$$J(u) = \int_0^T (u'^2 + u^2 + 2g(u)) \, dt. \qquad (8.5.2)$$

Let us assume that

(a) $|h(u)| \leq k_1 u^2$ as $u \to 0$,

(b) $|h(u) - h(v)| \leq k_2 |u - v|$, with $k_2 \ll 1$ for $|u|, |v| \ll 1$,

(c) $|b| \ll 1$.

Under these conditions, we will demonstrate that (8.5.1) possesses a solution for any $T > 0$. This solution is unique in the class of functions satisfying $|u| \leq 2|b|$ for $0 \leq t \leq T$. If (8.5.1) is derived from (8.5.2), then we know that the solution is unconditionally unique and yields the minimum of the convex functional $J(u)$.

We will further demonstrate that $u \equiv u(t, T)$ possesses a limit as $T \to \infty$ and that this limit satisfies the equation

$$u'' - u = h(u),$$

$$u(0) = b, \qquad \lim_{t \to \infty} u = 0. \qquad (8.5.4)$$

Observe that this equation is not covered by the Poincaré-Lyapunov theorem cited above since the associated characteristic roots are ± 1.

8.6.   The Associated Green's Function

Since our proof depends heavily upon some specific properties of the Green's function of

$$v'' - v = f, \qquad v(0) = 0, \qquad v'(T) = 0, \tag{8.6.1}$$

let us establish the results we need first for the sake of completeness. To begin with, we consider two initial-value problems. The first is

$$u'' - u = 0, \qquad u(0) = b, \qquad u'(T) = 0, \tag{8.6.2}$$

whose solution is readily seen to be

$$u = bc(T - t)/c(T), \tag{8.6.3}$$

where to simplify the subsequent notation we have set

$$s(t) = \sinh t = (e^t - e^{-t})/2,$$
$$c(t) = \cosh t = (e^t + e^{-t})/2. \tag{8.6.4}$$

For large T we have

$$u(t) = b(e^{-t} + e^{t-2T})/(1 + 0(e^{-T})) \cong be^{-t}, \tag{8.6.5}$$

for $0 \le t \le T$.

We next consider the equation

$$v'' - v = f, \qquad v(0) = 0, \qquad v'(T) = 0. \tag{8.6.6}$$

Its solution may be written

$$v = c_1 s(t) + \int_0^t s(t - t_1) f(t_1) \, dt_1. \tag{8.6.7}$$

The parameter c_1 is determined by the condition

$$v'(T) = 0. \tag{8.6.8}$$

Thus,

$$c_1 = -\int_0^T \frac{c(t - t_1) f(t_1)}{c(T)} \, dt_1, \tag{8.6.9}$$

while v takes the form

$$v = -\int_0^t \left[\frac{c(T - t_1) s(t)}{c(T)} - s(t - t_1) \right] f(t_1) \, dt_1 - \int_t^T \frac{c(T - t_1) s(t) f(t_1) \, dt_1}{c(T)}. \tag{8.6.10}$$

To simplify the term in the bracket, we use the identity

$$s(t - t_1) c(T) - c(T - t_1) s(t) = c(T - t) s(t_1). \tag{8.6.11}$$

This is readily seen by direct calculation, but it is more easily and elegantly demonstrated by noting that both sides are solutions of $u'' - u = 0$ with the same initial conditions as $t = T$. In any case, we anticipate the result because we know from other considerations that the Green's function is symmetric in t and t_1.

Hence, we may write

$$v = \int_0^T k(t, t_1) f(t_1) \, dt_1, \tag{8.6.12}$$

where $k(t, t_1)$, the Green's function for this set of boundary conditions, has the expression

$$k(t, t_1) \equiv k(t, t_1, T) = -\frac{c(T - t)s(t_1)}{c(T)}, \qquad 0 \le t \le t_1,$$

$$= -\frac{c(T - t_1)s(t)}{c(T)}, \qquad t_1 \le t \le T. \tag{8.6.13}$$

It follows from this that $k \le 0$ for $0 \le t, t_1 \le T$. However, we make no significant use of this basic property which can be established without use of a specific representation for $k(t, t_1, T)$. Using this result, however, we can readily obtain a useful estimate for the expression

$$w(t) = \int_0^T k(t, t_1, T) \, dt_1 \tag{8.6.14}$$

without any calculation. We observe that w is the solution of

$$w'' - w = 1, \qquad w(0) = 0, \qquad w'(T) = 0. \tag{8.6.15}$$

Set $w = -1 + v$, so that v satisfies

$$v'' - v = 0, \qquad v(0) = 1, \qquad v'(T) = 0, \tag{8.6.16}$$

whence

$$v = c(T - t)/c(T). \tag{8.6.17}$$

Hence, for $0 \le t_1 \le T$,

$$0 \le v \le 1, \qquad -1 \le w \le 0. \tag{8.6.18}$$

Thus,

$$\left| \int_0^T k(t, t_1, T) \, dt_1 \right| \le 1, \tag{8.6.19}$$

for $0 \le t_1 \le T$, $\quad T \ge 0$.

EXERCISES

1. Using the explicit expression for $k(t, t_1, T)$ show that $k(t, t_1, T) \sim e^{-(t-t_1)}$, $0 \le t_1 \le t$, $k(t, t_1, T) \sim e^{-(t_1-t)}$, $t \le t_1 < \infty$ as $T \to \infty$ and that a constant k_3 exists such that $|k(t, t_1, T)| \le k_3 e^{-|t-t_1|}$ for $0 \le t, t_1 \le T < \infty$.

2. Hence, obtain directly an estimate of the form

$$\int_0^T |k(t, t_1, T)| \, dt_1 \le a_1, \qquad 0 \le t_1 \quad T < \infty.$$

3. Show that the solution of (8.5.2) has the asymptotic form $u \sim ce^{-t}$ as $T \to \infty$.

4. What is the connection between $\lim_{T \to \infty} u(t, T)$ and the solution of $u'' - u = 0$, $u(0) = c$, $u(\infty) = 0$.

8.7. Conversion to an Integral Equation

We now possess the necessary ingredients for a proof of the result stated in Section 8.5., The equation in (8.5.1) can be converted into the integral equation

$$u \equiv u_0 + \int_0^T k(t, t_1)h(u) \, dt_1, \tag{8.7.1}$$

where

$$u_0 = bc(T - t)/c(T). \tag{8.7.2}$$

We write $k(t, t_1)$, omitting the variable T, to simplify the typography a bit.

To establish the existence of a solution of (8.7.1), we employ a prosaic method of successive approximations, setting

$$u_{n+1} = u_0 + \int_0^T k(t, t_1)h(u_n) \, dt_1 \qquad n \ge 0. \tag{8.7.3}$$

The proof of the convergence of the sequence $\{u_n\}$ proceeds in the usual fashion. We establish first uniform boundedness of $\{u_n\}$ and then convergence of $\sum_n |u_{n+1} - u_n|$ for $0 \le t \le T$.

We begin by establishing the result

$$|u_n| \le 2|c|, \tag{8.7.4}$$

for $|c| \ll 1$ and $0 \le t \le T$, inductively. The result is true for $n = 0$, since $|u_0| \le c$. From (8.7.3), we have

$$|u_{n+1}| \le |u_0| + \int_0^T |k(t, t_1)| \, |h(u_n)| \, dt_1$$

$$\le |c| + \int_0^T |k(t, t_1)|(k_1 |u_n|^2) \, dt_1, \tag{8.7.5}$$

upon referring to (8.5.3), Hence,

$$|u_{n+1}| \le |c| + 4k_1 c^2 \int_0^T |k(t, t_1)| \, dt_1$$

$$\le |c| + 4k_1 c^2 \le 2|c|, \tag{8.7.6}$$

for $|c| \le 1/4 \, k_1$.

To establish convergence, we write

$$u_{n+1} - u_n = \int_0^T k(t, t_1, T)(h(u_n) - h(u_{n-1})) \, dt_1,$$

$$\|u_{n+1} - u_n\| \le \|h(u_n) - h(u_{n-1})\| \int_0^T |k(t, t_1)| \, dt_1$$

$$\le k_2 \|u_n - u_{n-1}\|. \tag{8.7.7}$$

By assumption [see (8.5.3)], $k_2 < 1$ for $|u_n|$, $|u_{n-1}| \ll 1$. Hence the series $\sum_n \|u_{n+1} - u_n\|$ converges uniformly for $0 \le t \le T$, and the limit function $u = \lim_{n \to \infty} u_n$ satisfies (8.7.1).

8.8.   Conditional Uniqueness

A similar argument shows that this function is the unique solution in the class of functions satisfying the bound $|u| \le 2|c|$. Let v be another solution of (8.8.1) in the class. Then

$$u - v = \int_0^T k(t, t_1)(h(u) - h(v)) \, dt_1, \tag{8.8.1}$$

whence

$$\|u - v\| \le \|h(u) - h(v)\| \int_0^T |k(t, t_1)| \, dt_1$$

$$\le \|h(u) - h(v)\| \le k_2 \|u - v\|. \tag{8.8.2}$$

Hence, $\|u - v\| = 0$, $u \equiv v$.

EXERCISES

1. If $h(u) = u^2$, how large can we take c and still carry through the preceding proof ?
2. Is a restriction on c essential, or merely a consequence of the method we employ ? (*Hint:* $u'' - u = u^2$, $u(0) = c$, $u'(T) = 0$ can be solved explicitly.)

8.9.   Asymptotic Behavior of u—I

As a first step in obtaining the asymptotic behavior of $u(t, T)$ as $T \to \infty$, let us demonstrate that

$$|u_n| \le 3|c|e^{-t} \tag{8.9.1}$$

for $t \ge 0$. The result is certainly true for $n = 0$, since $|u_0| \le 2|c|e^{-t}$. We have, using (8.7.5),

$$|u_{n+1}| \le 2|c|e^{-t} + \int_0^T |k(t, t_1)|(k_1 u_n^2)\, dt_1$$

$$\le 2|c|e^{-t} + 9k_1|c|^2 \int_0^T |k(t, t_1)| e^{-2t_1}\, dt_1$$

$$\le 2|c|e^{-t} + 9k_1|c|^2 \int_0^T k_3 e^{-|t-t_1|} e^{-2t_1}\, dt_1$$

$$\le 3|c|e^{-t} \tag{8.9.2}$$

for $|c| \ll 1$. Consequently, the limit function $u(t)$ satisfies the same bound,

$$|u| \le 3|c|e^{-t}. \tag{8.9.3}$$

8.10.   Asymptotic Behavior

Using this result, we can show that $u(t, T)$ converges as $T \to \infty$ to a function $v(t)$ satisfying the equation

$$v'' - v = h(v), \qquad v(0) = c, \qquad \lim_{t \to \infty} v = 0. \tag{8.10.1}$$

To see this, consider the equivalent integral equation

$$v = ce^{-t} + \int_0^\infty k(t, t_1, \infty)h(v)\, dt_1, \tag{8.10.2}$$

where

$$k(t, t_1, \infty) = \lim_{T \to \infty} k(t, t_1, T). \tag{8.10.3}$$

The same argument that we have just used shows that there exists a solution of this equation for $|c| \ll 1$ and that it is unique in the class of functions for which $|v| \le 2|c|$.

We have

$$u - v = u_0 - ce^{-t} + \int_0^T k(t, t_1)h(u)\, dt_1 - \int_0^\infty k(t, t_1, \infty)h(v)\, dt_1$$

$$= u_0 - ce^{-t} + \int_0^T k(t, t_1)h(u)\, dt_1 - \int_0^T k(t, t_1, \infty)h(u)\, dt_1$$

$$- \int_T^\infty k(t, t_1, \infty)h(v)\, dt_1. \quad (8.10.4)$$

Hence,

$$|u - v| \le |u_0 - ce^{-t}| + \left| \int_0^T (k(t, t_1) - k(t, t_1, \infty))h(u)\, dt_1 \right|$$

$$+ \left| \int_0^T k(t, t_1, \infty)(h(u) - h(v))\, dt_1 \right| + \left| \int_T^\infty k(t, t_1, \infty)h(v)\, dt_1 \right|.$$

$$(8.10.5)$$

It is clear that

$$|u_0 - ce^{-t}| \le \varepsilon, \quad T \ge T_0. \quad (8.10.6)$$

Furthermore,

$$\left| \int_T^\infty k(t, t_1, \infty)h(v)\, dt_1 \right| \le k_4 \int_T^\infty e^{-|t - t_1|}\, dt_1 \le k_4 e^t \int_T^\infty e^{-t_1}\, dt_1 \le \varepsilon, \quad (8.10.7)$$

$T \ge T_0$.

The bound on u, $|u| \le 2|c|e^{-t}$, and thus $h(u)$, permits us to conclude that

$$\left| \int_0^T (k(t, t_1) - k(t, t_1, \infty))h(u)\, dt_1 \right| \le \varepsilon \quad (8.10.8)$$

for $T \ge T_0$.

Finally, we have

$$\left| \int_0^T k(t, t_1, \infty)(h(u) - h(v))\, dt_1 \right| \le \|h(u) - h(v)\| \int_0^T |k(t, t_1, \infty)|\, dt_1$$

$$\le k_5 \|u - v\|, \quad (8.10.9)$$

where $k_5 < 1$ for $|c| \ll 1$. Hence, (8.10.5) yields

$$\|u - v\| \le \frac{3\varepsilon}{1 - k_5} \quad (8.10.10)$$

for $T \ge T_0$. Thus,

$$\lim_{T \to \infty} u(t, T) = v(t). \quad (8.10.11)$$

EXERCISE

1.   For the case where $h(u) = u^2$, obtain a more precise bound for

$|u(t, T) - u(t, \infty)|$.

8.11.   Infinite Control Process

The foregoing results enable us to conclude that

$$\lim_{T \to \infty} \min_u \left[\int_0^T [u'^2 + 2g(u)] \, dt \right] = \min_u \left[\int_0^\infty (u'^2 + 2g(u)) \, dt_1 \right] \quad (8.11.1)$$

in the case where

$$g(u) = \frac{u^2}{2} + h(u), \quad (8.11.2)$$

with $h(u)$ convex in addition to the previous requirements.

8.12.   Multidimensional Case

Let us now briefly turn to the equation

$$x'' - Ax = h(x), \qquad x(0) = c, \qquad x'(T) = 0, \quad (8.12.1)$$

associated with the functional

$$J(x) = \int_0^T [(x', x') + (x, Ax) + 2g(x)] \, dt. \quad (8.12.2)$$

There is no difficulty in obtaining the analogues of the results for the scalar case. The basic preliminary result is the analogue of (8.6.12). Let A be positive definite and let B be the positive definite square root of A. Introduce the matrix functions

$$S(t) = (e^{Bt} - e^{-Bt})/2,$$
$$C(t) = (e^{Bt} + e^{-Bt})/2. \quad (8.12.3)$$

Since $S(t)$ and $C(t)$ commute with each other and with the functional values obtained for different values of t, we readily obtain the results corresponding to (8.6.2) and (8.6.12). From there on the details are the same, apart from a precise estimate such as (8.6.19). This must be replaced by the result indicated in Exercise 2 at the end of Section 8.6.

8.13. Asymptotic Control

In the previous sections, we discussed the asymptotic nature of the minimum of

$$J(x) = \int_0^T [(x', x') + (x, Ax) + 2g(x)] \, dt \tag{8.13.1}$$

under the assumption that $\|x(0)\| \ll 1$. Let us now show that if we add some further reasonable assumptions concerning $g(x)$, we can allow an arbitrary initial condition. Precisely, we wish to show that if we assume

(a) $A > 0,$
(b) $g(x) \geq 0,$
(c) $g(x)$ is convex,   (8.13.2)
(d) $\int_0^\infty g(ce^{-t}) \, dt < \infty$ for any c,

then we can assert that $\|x(t, T)\| \to 0$ as $t, T \to \infty$. Once we have this result we can readily obtain the asymptotic behavior more precisely. Here $x(t, T)$, in our usual notation, is the function minimizing $J(x)$. Observe that there is now no condition on the magnitude of $\|c\|$, where c is the initial state.

The proof consists of a series of simple individual results.

8.14. **Boundedness of** $\|x(t)\|$

Let us begin by establishing the uniform boundedness of $x(t)$. We have

$$J(x) = \int_0^T [(x', x') + (x, Ax) + 2g(x)] \, dt \leq J(x_0) \tag{8.14.1}$$

for any admissible function $x_0(t)$. Choose $x_0 = ce^{-t}$. Then

$$J(x_0) = \int_0^T [e^{-t}\{(c, c) + (c, Ac)\} + 2g(ce^{-t})] \, dt < \int_0^\infty [\cdots] \, dt = g_1(c). \tag{8.14.2}$$

Hence,

$$|(x(t), x(t)) - (c, c)| = \left| 2 \int_0^t (x', x) \, dt_1 \right| \leq \int_0^t [(x, x) + (x', x')] \, dt_1$$

$$\leq k_2 \int_0^t [(x, Ax) + (x', x')] \, dt_1$$

$$\leq k_2 \int_0^t [(x, Ax) + (x', x') + 2g(x)] \, dt_1 \leq k_2 g_1(c), \tag{8.14.3}$$

where k_2 is a constant determined by A. Here we have used the Cauchy-Schwarz inequality, the positive definite character of A and the nonnegativity of $h(x)$.

EXERCISE

1. Can we replace $g(x) \geq 0$ by $g(x) \geq -(x, Cx)$ for suitable C?

8.15. Convergence of $f(c, T)$

We have already indicated that the foregoing hypotheses lead to the convergence of

$$f(c, T) = \min_x J(x). \tag{8.15.1}$$

Hence,

$$f(c, T) - f(c, S) < \varepsilon \tag{8.15.2}$$

for $T > S \geq T_0(\varepsilon)$. From this we may conclude that

$$\int_0^T \phi(x(t, T)) \, dt - \int_0^S \phi(x(t, S)) \, dt$$

$$= \int_0^S \phi(x(t, T)) \, dt - \int_0^S \phi(x(t, S)) \, dt + \int_S^T \phi(x(t, T)) \, dt < \varepsilon,$$

where

$$\varphi(x) = (x', x') + (x, Ax) + 2g(x), \tag{8.15.4}$$

the integrand in $J(x)$.

Since $x(t, S)$ is the minimizing function over $[0, S]$, we have

$$\int_0^S \phi(x(t, T) \, dt \geq \int_0^S \phi(x(t, S)) \, dt. \tag{8.15.5}$$

Hence, (8.15.3) yields

$$\int_S^T \phi(x(t, T)) \, dt \leq \varepsilon, \tag{8.15.6}$$

for $T \geq S \geq T_0(\varepsilon)$.

From the nonnegativity of $h(x)$ we can conclude from (8.15.6) that

$$\int_S^T (x', x') \, dt \leq \varepsilon, \qquad \int_S^T (x, Ax) \, dt \leq \varepsilon. \tag{8.15.7}$$

Using again the fact that A is positive definite, this yields

$$\int_S^T (x, x)\, dt \le k_1 \varepsilon. \tag{8.15.8}$$

8.16. Conclusion of Proof

From the foregoing it is easy to conclude that $\|x(t)\|$ becomes small as $t, T \to \infty$. Take $T - S > 1$. The first inequality in (8.15.7) tells us that there must be a point t_2 in $[S, T]$ where $(x, x) \le \varepsilon$. Consider then the expression

$$(x(t), x(t)) - (x(t_2), x(t_2)) = 2 \int_{t_2}^t (x, x')\, dt_1, \tag{8.16.1}$$

for $T \ge t \ge t_2 \ge S$. We have

$$|(x(t), x(t)) - (x(t_2), x(t_2))| \le \int_{t_2}^t [(x, x) + (x', x')]\, dt_1 \le \varepsilon + k_1 \varepsilon. \tag{8.16.2}$$

Thus, once $x(t)$ becomes small, as it must, it remains small.

8.17. Infinite Processes

Consider the function $f(c, T)$ determined by

$$f(c, T) = \min_u J(u) = \min_u \left[\int_0^T (u'^2 + 2g(u))\, dt \right], \tag{8.17.1}$$

where $u(0) = c$ and $0 \le T < \infty$. In the previous sections we considered some results which enable us to conclude that $f(c, T)$ possesses a limit as $T \to \infty$ and, in some cases, that

$$\lim_{T \to \infty} f(c, T) = \min_u \left[\int_0^\infty (u'^2 + 2g(u)) \right] dt, \tag{8.17.2}$$

with a corresponding relation in the multidimensional case. Let us now consider the infinite process directly, proceeding in a formal fashion. Write

$$f(c) = \min_u \left[\int_0^\infty (u'^2 + 2g(u))\, dt \right]. \tag{8.17.3}$$

Then, as before,

$$f(c) = \min_v [(v^2 + 2g(c))\Delta + f(c + v\Delta)] + o(\Delta),$$

$$0 = \min_v [(v^2 + 2g(c)) + vf'(c)],$$

$$0 = 2g(c) - f'(c)^2/4, \tag{8.17.4}$$

Similarly, in the multidimensional case, if we write

$$f(c) = \min_{x} \left[\int_0^{\infty} [(x', x') + 2g(x)] \, dt \right], \tag{8.17.5}$$

$x(0) = c$, we obtain formally

$$0 = 2g(c) - (\text{grad } f, \text{grad } f)/4. \tag{8.17.6}$$

When results of this nature can be established rigorously, they can be of considerable use in furnishing approximations to $f(c, T)$, $v(c, T)$, and $u(c, T)$ for $T \gg 1$.

EXERCISES

1. Consider the case where $f(c) = \min_u \int_0^{\infty} (u'^2 + u^2) \, dt$, $u(0) = c$, and show that (8.17.4) is valid.

2. Show that (8.17.4) leads to

$$f'(c) = \min_{v \le 0} \left[-\left(\frac{v^2 + 2g(c)}{v} \right) \right]$$

if $g(c) > 0$. What is an economic meaning of this?

3. Obtain the corresponding relations for the cases where

   (a) $J(u) = \int_0^{\infty} (u'^2 + e^{-bt}u^2) \, dt,$ $u(0) = c,$

   (b) $J(u, v) = \int_0^{\infty} e^{-bt}g(u - v) \, dt,$ $u' = h(v),$ $u(0) = c,$

$$0 < v < u.$$

4. If $g(c) = c^2 + a_2 c^4 + \cdots$, show that (8.17.4) permits the determination of $f(c)$ as a power series in c^2.

5. Under what conditions can one establish (8.17.4)?

6. Show that $(c, A^{1/2}c) = \min_x [\int_0^{\infty} [(x', x') + (x, Ax)] \, dt]$ if $x(0) = c$, $A > 0$.

7. If $g(x) = (x, Ax) + \cdots$, show that (8.17.6) permits the determination of the power series expansion of $f(c)$.

8.18.   Discrete Infinite Processes

Let us return to a discrete process

$$x_{n+1} = g(x_n, y_n), \qquad x_0 = c,$$

$$J(\{x_n, y_n\}) = \sum_{n=0}^{N} h(x_n, y_n). \tag{8.18.1}$$

Then, setting

$$f_N(c) = \min_{\{y_n\}} J(\{x_n, y_n\}), \tag{8.18.2}$$

we obtain the functional equation

$$f_N(c) = \min_y [h(c, y) + f_{N-1}(g(c, y))], \tag{8.18.3}$$

$N \geq 1$, $f_0(c) = \min_y h(c, y)$. Letting $N \to \infty$, we formally obtain the functional equation

$$f(c) = \min_y [h(c, y) + f(g(c, y))]. \tag{8.18.4}$$

There are a number of interesting questions concerning the existence and uniqueness of solutions of equations of the type appearing above, many of significant difficulty. We shall not, however, discuss any of them here since they belong more properly to the theory of dynamic programming than to control theory.

EXERCISE

1. The functional equation $u(c) = \min_v [c^2 + v^2 + u(c + bv)]$ possesses a solution of the form $u(c) = kc^2$. Does it possess any other strictly convex solutions? (See

   R. Bellman, "Functional Equations in the Theory of Dynamic Programming—XVI: An Equation Involving Iterates," *JMAA*, to appear.

8.19. Steady-state Average Behavior

Let us next consider a type of asymptotic behavior quite different from that encountered in the previous section. Consider a functional equation of the form

$$f_n(i) = \max_q [a(i, q) + f_{n-1}(T(i, q))], \qquad n \geq 1,$$
$$f_0(i) = \max_q [a(i, q)], \tag{8.19.1}$$

where $i = 1, 2, \ldots, N$, and q ranges over a finite set of values such that $T(i, q)$ is again an integer in the set $1, 2, \ldots, N$. In a number of cases, the economic or engineering background leads us to surmise that there exists an asymptotic relation of the form

$$f_n(i) \sim a_1 n \tag{8.19.2}$$

as $n \to \infty$, where a_1 is independent of i. The quantity a_1 then represents the maximum average gain.

We wish to demonstrate that this important result can be readily obtained under simple and reasonable assumptions. These are

(a) $0 \leq a(i, q) < \infty$,
(b) $T(i, q)$ is such that by means of a suitable succession of decisions $q_1, q_2, \ldots, q_k$, it is possible to go from any state i to any other state j.

This hypothesis concerning T is of an *ergodic* nature.

8.20.   Subadditive Functions

The proof hinges on a lemma due to Fekete which seems specifically designed for multistage decision processes of this nature.

LEMMA.   *Let $\{u_n\}$ be a nonnegative sequence such that*

$$u_{m+n} \leq u_m + u_n, \quad n = 1, 2, \ldots, \tag{8.20.1}$$

Then

$$\lim_{n \to \infty} u_n/n = a_1. \tag{8.20.2}$$

Proof.   Since $u_n \leq nu_1$ by virtue of (8.20.1), it is clear that u_n/n is uniformly bounded. Let

$$a_1 = \underline{\lim}\, u_n/n, \tag{8.20.3}$$

and let N be such that

$$a_1 \cong u_N/N. \tag{8.20.4}$$

Let us then show that

$$\lim_{R \to \infty} U_{RN}/RN = a_1. \tag{8.20.5}$$

This follows from (8.20.1) and the definition of a_1, since

$$\frac{U_{RN}}{RN} \leq \frac{RU_N}{RN} = \frac{U_N}{N}, \tag{8.20.6}$$

for $R = 1, 2, \ldots$, and a_1 is the lim inf.

Let $M = RN + S$, $0 \leq S \leq N - 1$. Then

$$\frac{U_M}{M} = \frac{U_{RN+S}}{RN + S} \leq \frac{U_{RN} + U_S}{RN + S} \leq \frac{U_{RN}}{RN}\left(\frac{RN}{RN + S}\right) + \frac{U_S}{RN + S}. \tag{8.20.7}$$

If we let M be a value such that

$$\frac{U_M}{M} \cong \overline{\lim}\, \frac{u_n}{n} = a_2, \tag{8.20.8}$$

we obtain from (8.20.7)

$$a_2 \leq a_1 \left(\frac{RN}{RN + S} \right) + \frac{U_S}{RN + S} + \varepsilon, \qquad (8.20.9)$$

with ε as small as desired for M sufficiently large. Take a sequence of increasing M's, which is to say, let $R \to \infty$, and we obtain

$$a_2 \leq a_1, \qquad (8.20.10)$$

whence $a_2 = a_1$. This means the stated limit exists.

EXERCISES

1. What happens if we remove the nonnegativity constraint on the sequence $\{u_n\}$?
2. Let $f(n)$ be defined for $n = 1, 2, \ldots$, satisfy $f(mn) = f(m)f(n)$ for m and n relatively prime and be monotone increasing; then $f(n) \sim k \log n$ as $n \to \infty$. (Erdos)

8.21. Proof of Theorem

Introduce the new sequence

$$u_n = \max_i f_n(i). \qquad (8.12.1)$$

It follows from the definition of the multistage process that

$$u_{m+n} \leq u_m + u_n. \qquad (8.12.2)$$

From the lemma in Section 8.20 it follows that $u_n/n \sim a_1$ as $n \to \infty$.

For each n, let i_n be a value of i for which $f_n(i)$ achieves its maximum value. Let $q_1, q_2, \ldots, q_k$ be a sequence of decisions which transforms i into i_{n-M} where M is the maximum number of decisions required to transform any point i into any other point j.

Then, since the $a(i, q)$ are nonnegative,

$$f_n(i) \geq f_{n-k}(i_{n-M}). \qquad (8.21.3)$$

Furthermore,

$$f_{n-k}(i_{n-M}) \geq f_{n-M}(i_{n-M}), \qquad (8.21.4)$$

where we take $n > M$. Using the obvious relation $f_n(i_n) \geq f_n(i)$, we have, for large n,

$$n(a_1 + \varepsilon) \geq f_n(i_n) \geq f_n(i) \geq f_{n-M}(i_{n-M}) \geq (n - M)(a_1 - \varepsilon). \qquad (8.21.5)$$

It follows that

$$f_n(i) \sim na_1,\qquad(8.21.6)$$

as $n \to \infty$, where a_1 is independent of the initial state i.

EXERCISES

1. Let the functional equation be $f_n(p) = \max_q [a(p, q) + f_{n-1}(T(p, q))]$, $p \in R$. What conditions on the continuous transformation $T(p, q)$ will yield a corresponding result?

2. If $u_{m+n}(p) \le u_m(p) + u_n(T(p))$, for p, $T(p) \in R$, under what additional conditions will $u_n(p)/p \to a_1$?

3. Let $p \in R$, $T(p) \in R$ where T is continuous and R is finite. Under what additional conditions will

$$\frac{u_n(p)}{n+1} = \frac{p + T(p) + T^{(2)}(p) + \cdots + T^{(n)}(p)}{(n+1)}$$

approach a limit? Here $T^{(n)} = T(T^{(n-1)})$, $n \ge 1$.

Miscellaneous Exercises

1. Consider the Riccati equation $u' = u^2 - t^2 + 1 + a - 2bt$. Let $u = u(t)$ be the solution determined for a particular $a > 0$. Then there is a unique $b_0 = b_0(a)$ with the properties that

   (a) If $b > b_0$, then $u \to -\infty$ as $t \to +\infty$;
   (b) If $b < b_0$, then $u \to +\infty$ at a vertical asymptote $t = t_0(a, b)$;
   (c) If $b = b_0$, then $u \ge 0$ in $0 \le t < \infty$ and $u = t + b + 0(1)$ as $t \to +\infty$.

   Furthermore, $b_0(a)$ is a continuous monotone increasing function of a.

   (J. E. Littlewood, "On Nonlinear Differential Equations of the Second Order—I and II," *Acta Math.* **97**, 1957, pp. 267–308; *Acta Math.*, **98**, 1957, pp. 1–110.)

2. Show that $b_0(a) \sim a^{3/2}$ as $a \to \infty$.

   (H. P. F. Swinnerton-Dyer, "On a Problem of Littlewood Concerning Riccati's Equation," *Proc. Camb. Phil. Soc.*, **65**, 1969, pp. 651–652.)

3. Let
   $$f_1(q) = aq + b, \quad q = 1, 2, \ldots,$$
   $$f_k(q) = \max_{l=1, 2, \ldots, q}[a(k)l + b(k) f_{k-1}(l)] \cdot q + c(k)], \qquad k \ge 2.$$
   Determine the asymptotic behavior of $f_k(q)$ as $q \to \infty$.

   (P. Wegner, "The Asymptotic Properties of a Class of Dynamic Programming Formulas," *J. Math. Anal. Appl.*, **12**, 1965, pp. 570–575.)

BIBLIOGRAPHY AND COMMENTS

8.1.   For a discussion of classical results concerning the solution of $x' = Ax + h(x)$, $x(0) = c$, the theory of Poincaré and Lyapunov, and references to additional results, see

R. Bellman, *Methods of Nonlinear Analysis*, Academic Press, New York, 1970.
R. Bellman, *Stability Theory of Differential Equations*, Dover, New York, 1970.

For some results concerning the asymptotic behavior of solutions of functional equations of the form

$$f_n(p) = \max_q [g(p, q) + f_{n-1}(T(p, q))],$$

as $n \to \infty$, see

R. Bellman, *Dynamic Programming*, Princeton Univ. Press, Princeton, N. Jersey, 1957.
8.5.   See

R. Bellman, "On Analogues of Poincaré-Lyapunov Theory for Multipoint Boundary-value Problems—I, "*J. Math. Anal. Appl.*, **14**, 1966, pp. 522–526.

R. Bellman, "On Analogues of Poincaré-Lyapunov Theory for Multipoint Boundary-value Problems—Correction," *J. Math. Anal. Appl.* to appear.
8.6.   For further results concerning Green's functions, see

R. Bellman, *Introduction to the Mathematical Theory of Control Processes*, Vol. I: *Linear Equations and Quadratic Criteria*, Academic Press, New York, 1967.
8.8.   See

R. Bellman and R. Bucy, "Asymptotic Control Theory," *J. SIAM Control*, **2**, 1964, pp 11–18.
8.13.   See

R. Bellman, "A Note on Asymptotic Control Theory," *J. Math. Anal. Appl.*, to appear.
8.20.   The representation for the square root of A was given in

R. Bellman, "Some Inequalities for the Square Root of a Positive Definite Matrix," *Linear Algebra Its Appl.* **1**, 1968, pp. 321–324.
8.21.   See the book on dynamic programming cited in 8.1.
8.22.   See

R. Bellman, "Functional Equations in the Theory of Dynamic Programming—XI: Limit Theorems," *Rend. Circ. Mat. Palermo*, **8**, 1959, pp. 1–3.
8.23.   A special case is due to Fekete. The general case is due to Polya and Szego. See

G. Polya and G. Szego, *Aufgaben und Lehrsatze aus der Analysis*, Dover, New York, 1945, p. 171) solution to 98).

9

——

DUALITY AND UPPER AND LOWER BOUNDS

9.1. Introduction

Consider the by now familiar problem of minimizing the functional

$$J(u) = \int_0^T (u'^2 + 2g(u)) \, dt, \qquad (9.1.1)$$

subject to the initial condition $u(0) = c$. As we know, there is no difficulty in obtaining upper bounds for $\min_u J(u)$. For any admissible function v we have

$$\min_u J(u) \le J(v). \qquad (9.1.2)$$

We have previously discussed some techniques for obtaining numerical estimates of both the minimum value and the minimizing function under various conditions that assure that the minimum is assumed. All of these methods would be tremendously enhanced if we could devise some equally systematic procedures for the derivation of *lower* bounds. As we have pointed out, one of the major hurdles in the application of the Bubnov-Galerkin and Raleigh-Ritz methods is the determination of how much effort is required to obtain a desired accuracy.

Since the values of the minimizing function usually constitute a "flat" surface in function space, it is often difficult to obtain any significant improvements of the values obtained by quite simple initial approximations. This is a well-known observation which does not seem to have been analyzed in any precise terms.

Fortunately, a number of procedures exist for finding upper and lower bounds, all of which center about the fundamental theme of *duality*. We will present some results which illustrate the general concept and which can be obtained using reasonably simple tools. More sophisticated methods centering about convexity are needed for more difficult optimization processes involving constraints.

9.2. Formalism

Let us begin by presenting an example of the basic formalism. We start with the observation that

$$u'^2 = \max_v [2u'v - v^2]. \tag{9.2.1}$$

This permits us to write

$$\int_0^T (u'^2 + 2g(u)) \, dt = \max_v \left[\int_0^T (2u'v - v^2 + 2g(u)) \, dt \right]. \tag{9.2.2}$$

Hence, for any function $v(t)$ in $L^2(0, T)$ we have

$$\int_0^T (u'^2 + 2g(u)) \, dt \geq \left[\int_0^T (2u'v - v^2 + 2g(u)) \, dt \right]. \tag{9.2.3}$$

Thus,

$$\min_u \left[\int_0^T (u'^2 + 2g(u)) \, dt \right] \geq \min_u \left[\int_0^T (2u'v - v^2 + 2g(u)) \, dt \right], \tag{9.2.4}$$

valid for any function $v(t)$ in $L^2(0, T)$. Let us further restrict the class of v's by requiring that v' exists. Since we are interested in obtaining bounds we can choose v to our own satisfaction.

To carry through the minimization with respect to u on the right, we begin by integrating by parts,

$$\int_0^T 2u'v \, dt = 2uv \Big]_0^T - \int_0^T 2uv' \, dt. \tag{9.2.5}$$

Since equality in (9.2.1) is attained for $v = u'$ and since we know that the minimization of $J(u)$ imposes the additional constraint $u'(T) = 0$, we restrict our attention to v-functions satisfying the condition $v(T) = 0$. Hence, (9.2.5) yields

$$\int_0^T 2u'v \, dt = -2u(0)v(0) - \int_0^T 2uv' \, dt$$
$$= -2cv(0) - \int_0^T 2uv' \, dt. \tag{9.2.6}$$

Using this result, the right-hand side of (9.2.4) becomes

$$\min_{u} \left[-2cv(0) + \int_0^T (2g(u) - 2uv') \, dt - \int_0^T v^2 \, dt \right]. \qquad (9.2.7)$$

We now introduce the Legendre-Fenchel function

$$h(w) = \min_{u} (g(u) - uw), \qquad (9.2.8)$$

of which we will say more later. Carrying out the minimization with respect to u in (9.2.7) then yields the lower bound,

$$\min_{u} J(u) \geq \left[-2cv(0) + \int_0^T (2h(v') - v^2) \, dt \right], \qquad (9.2.9)$$

for all v satisfying the constraint

$$v(T) = 0, \qquad (9.2.10)$$

and such that $\int_0^T h(v') \, dt$ exists. Let us introduce the new functional

$$K(v) = \left[-2cv(0) + \int_0^T (2h(v') - v^2) \, dt \right]. \qquad (9.2.11)$$

Then, we can write

$$\min_{u} J(u) \geq \max_{v} K(v), \qquad (9.2.12)$$

where the maximization is taken over functions of the foregoing nature. This furnishes another lower bound, which is the best of this type.

What we will demonstrate below is that under the assumption of convexity of $g(u)$, we actually have the equality

$$\min_{u} J(u) = \max_{v} K(v), \qquad (9.2.13)$$

where u and v, respectively, are constrained by the end conditions $u(0) = c$ and $v(T) = 0$, and restriction to appropriate function classes. The foregoing lower bound is thus the best lower bound available, namely the value of the minimum itself. It is interesting, however, to note that (9.2.13) is not necessary to make the method useful. The numerical application of (9.2.13) will be discussed in Section 9.5.

9.3. Quadratic Case

The proofs we present in the following pages, which are simple and direct, are based upon the explicit calculation and comparison of both sides of

(9.2.13). Let us begin with the easiest case, that where the function g is quadratic. We have

$$J(u) = \int_0^T (u'^2 + u^2)\, dt \geq \max_v \left[-2cv(0) - \int_0^T (v'^2 + v^2)\, dt \right], \quad (9.3.1)$$

upon applying the foregoing results.

The Euler equation associated with this new variational problem posed by the right-hand side is

$$v'' - v = 0, \qquad v'(0) = c, \qquad v(T) = 0. \quad (9.3.2)$$

The Euler equation associated with the minimization of $J(u)$ is

$$u'' - u = 0, \qquad u(0) = c, \qquad u'(T) = 0. \quad (9.3.3)$$

Comparing these two equations, we observe that $v = u'$ is a solution of (9.3.2). We have the advantage, of course, of suspecting that this had to be the case from the very beginning. Let us verify the details. We have

$$v' = u'' = u,$$
$$v'' = u' = v, \quad (9.3.4)$$

showing that the differential equation is satisfied. Furthermore,

$$v'(0) = u(0) = c,$$
$$v(T) = u'(T) = 0, \quad (9.3.5)$$

showing that the boundary conditions hold.

The uniqueness of the solution of (9.3.2) (established in Volume I) yields the desired identity,

$$v = u'. \quad (9.3.6)$$

Finally, from (9.3.2) we have

$$\int_0^T v(v'' - v)\, dt = 0, \qquad vv' \Big]_0^T - \int_0^T (v'^2 + v^2)\, dt = 0. \quad (9.3.7)$$

Thus,

$$\max_v \left[-2cv(0) - \int_0^T (v'^2 + v^2)\, dt \right] = -cv(0) = cu'(0), \quad (9.3.8)$$

precisely the value of $\min_u J(u)$ obtained in Volume I. Thus, a direct calculation establishes the desired equality.

EXERCISES

1. Establish the uniqueness of the solution of (9.3.2) using the device in (9.3.7).

2.   Using the method of (9.3.5), show that $\min_u J(u) = cu'(0)$.

3.   Use the dual problem and the Rayleigh-Ritz method to obtain estimates of

(a) $\displaystyle\min_u \int_0^1 (u'^2 + u^2)\, dt, \qquad u(0) = 1,$

(b) $\displaystyle\min_u \int_0^\infty (u'^2 + u^2)\, dt, \qquad u(0) = 1,$

(c) $\displaystyle\min_u \int_0^1 (u'^2 + e^u)\, dt, \qquad u(0) = 1.$

9.4.   Multidimensional Quadratic Case

The foregoing arguments extend immediately to the vector case. Let the criterion functional have the form

$$J(x) = \int_0^T [(x', x') + (x, Ax)]\, dt, \qquad x(0) = c, \tag{9.4.1}$$

assuming that $A > 0$. This ensures that the minimum exists and is determined by the Euler equation for $T \geq 0$ and all c. The result we need, corresponding to (9.2.1) is

$$(x', x') = \max_y [2(x', y) - (y, y)]. \tag{9.4.2}$$

Thus,

$$J(x) \geq \int_0^T [2(x', y) - (y, y) + (x, Ax)]\, dt \tag{9.4.5}$$

for all y, such that $y \in L^2(0, T)$. We further suppose that y' exists and belongs to $L^2(0, T)$. Integrating by parts as before,

$$J(x) \geq \left[-2(c, y(0)) + \int_0^T [(x, Ax) - 2(x, y')]\, dt - \int_0^T (y, y)\, dt \right] \tag{9.4.4}$$

Thus,

$$\min_x J(x) \geq \min_x \left[\int_0^T \{(x, Ax) - 2(x, y')\}\, dt - 2(c, y(0)) - \int_0^T (y, y)\, dt \right] \tag{9.4.5}$$

Under the assumption that $A > 0$, we have

$$\min_x \{(x, Ax) - 2(x, y')\} = -(y', A^{-1}y') \tag{9.4.6}$$

as a direct calculation shows. Thus, we obtain the inequality

$$\min_x J(x) \geq \max_y \left[-2(c, y(0)) - \int_0^T [(y', A^{-1}y') + (y, y)] \, dt\right], \quad (9.4.7)$$

where $y(T) = 0$.

A direct paraphrase of the results of the preceding section shows that equality actually holds.

EXERCISE

1. Carry out the details.

9.5. Numerical Utilization

If we wish to obtain numerical results without treating the Euler equation, we can apply the Rayleigh-Ritz method to both functionals,

$$J(x) = \int_0^T [(x', x') + (x, Ax)] \, dt \qquad (9.5.1)$$

and

$$K(y) = -2(c, y(0)) - \int_0^T [(y', A^{-1}y') + (y, y)] \, dt. \qquad (9.5.2)$$

Since we know that $x' = y$ holds for the minimizing x and the maximizing y, if we choose the trial function,

$$x = \sum_{k=1}^{M} a_k w_k(t), \qquad (9.5.3)$$

in the first functional, with

(a) $\sum_{k=1}^{M} a_k w_k(0) = c,$

(b) $\sum_{k=1}^{M} a_k w_k'(T) = 0$ $\qquad (9.5.4)$

we can expeditiously choose the trial function

$$y = \sum_{k=1}^{M} a_k w_k'(t) \qquad (9.5.5)$$

in the second functional. On the other hand, it may be better in some cases to omit condition (9.5.4b) and use a different trial function for y.

One of the principal advantages of the Rayleigh-Ritz method in this case is that it may furnish a very desirable reduction of dimensionality in the case where N, the dimension of x, is large. (We have discussed dimensionality difficulties at length in Volume I.) Note that A^{-1} is required if we wish to employ the foregoing method. As we have previously emphasized, calculations of inverses can be difficult.

EXERCISE

1. How could one use (9.4.6) to obtain approximations to A^{-1}?

9.6. The Legendre-Fenchel Transform

Let us next present some auxiliary results we shall employ below. Consider the transform

$$h(w) = \min_{u} [g(u) - uw], \qquad (9.6.1)$$

which we call the *Legendre-Fenchel Transform*. We wish to present some of the basic properties of this operation.

Let us suppose that $g(u)$ is strictly convex, $g''(u) > 0$, for $-\infty < u < \infty$. Then a unique minimizing value exists in (9.6.1), given by the solution of

$$g'(u) - w = 0. \qquad (9.6.2)$$

Write $u = u(w)$ to denote the function determined in this fashion. Then,

$$h'(w) = (g'(u) - w)u'(w) - u = -u. \qquad (9.6.3)$$

From (9.6.2), we have

$$g''(u)u'(w) = 1, \qquad (9.6.4)$$

whence

$$u'(w) = \frac{1}{g''(u)} > 0. \qquad (9.6.5)$$

Returning to (9.6.3), we have

$$h''(w) = -u'(w) < 0. \qquad (9.6.6)$$

Hence, the convexity of g implies the concavity of h.

The symmetric nature of the foregoing results leads us to surmise that a simple inversion formula exists, namely

$$g(u) = \max_{w} (h(w) + uw). \qquad (9.6.7)$$

EXERCISES

1. Establish the concavity of h directly from the relation $h(\lambda w_1 + (1 - \lambda)w_2) = \min_u [g(u) - (\lambda w_1 + (1 - \lambda)w_2)u)]$, without calculation.

2. Determine the Legendre-Fenchel Transform of u^a, $a > 1$, and e^{bu}, $b > 0$.

3. What is the relation between the transform of $ag(u)$ and $ag(bu)$, $a, b > 0$?

4. If $g(u)$ is strictly convex, then $g(u) = \max_v [g(v) + (u + v)g'(v)]$.

5. Show that (9.2.1) is a special case of the preceding result.

9.7. Convex *g*

Let us now turn to the minimization of the functional

$$J(u) = \int_0^T (u'^2 + 2g(u))\, dt, \tag{9.7.1}$$

under the assumption that g is strictly convex in u. The Euler equation is

$$u'' - g'(u) = 0, \qquad u(0) = c, \qquad u'(T) = 0. \tag{9.7.2}$$

The companion functional obtained, as derived in Section 9.2, is

$$K(v) = -2cv((0 = \int_0^T (2h(v') - v^2)\, dt, \tag{9.7.3}$$

where h is the Legendre-Fenchel transform of g,

$$h(w) = \min_u (g(u) - uw). \tag{9.7.4}$$

The associated Euler equation is readily seen to be

$$[h'(v')]' + v = 0 \qquad h'(v')\Big|_{t=0} = c, \qquad v(T) = 0. \tag{9 7.5}$$

We know from the concavity of h that the solution exists and is unique, and we suspect from the derivation of $K(v)$ that it is very simply determined by the relation $v = u'$, with u the unique solution of (9.7.2). Let us verify this. Consider the function $z = u'(t)$. Then, using (9.7.2),

$$z' = u'' = g'(u). \tag{9.7.6}$$

Hence,

$$h'(z') = h'(g'(u)). \tag{9.7.7}$$

Let us now use the results of Section 9.6. From (9.7.4), we have

$$h'(w) = -u, \qquad h'(g'(u)) = -u. \tag{9.7.8}$$

Hence, (9.7.7) yields

$$h'(z') = -u, \tag{9.7.9}$$

whence

$$[h'(z')]' = -u' = -z. \tag{9.7.10}$$

This shows that u' satisfies the differential equation. The fact that the boundary conditions hold follows from (9.7.9) and the relation $v = u'$. Uniqueness completes the proof.

EXERCISE

1. How would one construct a dual functional in the case where

$$J(u) = \int_0^T g(u', u) \, dt,$$

with g convex in both arguments?

9.8.   Multidimensional Legendre-Fenchel Transform

In order to construct the corresponding proof of the equivalence of the dual processes for the multidimensional case where

$$J(x) = \int_0^T [(x', x') + 2g(x)] \, dt, \tag{9.8.1}$$

we require analogues of the results of Section 9.6. Let $h(y)$ be defined by

$$h(y) = \min_x [g(x) - (x, y)], \tag{9.8.2}$$

where $g(x)$ is strictly convex. By this we mean that the matrix

$$H(g) = (g_{ij}), \qquad g_{ij} = \frac{\partial^2 g}{\partial x_i \, \partial x_j} \tag{9.8.3}$$

is positive definite for all x. The matrix H is called the Hessian of g.

The minimum in (9.8.2) is then attained by the unique solution of the equation

$$\operatorname{grad} g(x) = y. \tag{9.8.4}$$

For x determined in this fashion, we have

$$\frac{\partial h(y)}{\partial y_i} = \sum_{j=1}^N g_j \frac{\partial x_j}{\partial y_i} - \sum_{j=1}^N \frac{\partial x_j}{\partial y_i} y_j - x_i = -x_i, \tag{9.8.5}$$

$i = 1, 2, \ldots, N$, where

$$g_j = \frac{\partial g(x)}{\partial x_j}. \tag{9.8.6}$$

Hence, we have the reciprocal formula

$$\text{grad } h(y) = -x. \tag{9.8.7}$$

From (9.8.7), we have

$$H(h)\left(\frac{\partial y_i}{\partial x_j}\right) = -I, \tag{9.8.8}$$

where I is the identity matrix. Turning to (9.8.4), we see that

$$H(g) = \frac{\partial y_i}{\partial x_j}. \tag{9.8.9}$$

Hence,

$$H(h)H(g) = -I. \tag{9.8.10}$$

Thus, if $H(g)$ is positive definite, $H(h)$ must be negative definite, which means that h is strictly concave.

The symmetric nature of the foregoing results leads to the inversion formula

$$g(x) = \max_{y} \, [h(y) + (x, y)]. \tag{9.8.11}$$

EXERCISES

1. Establish (9.8.11).

2. If $g(x)$ is strictly convex, then $g(x) = \max_y \, [g(y) + (J(y), x - y)]$, where $J(y)$ is the Jacobian of g evaluated at y.

9.9. Convex $g(x)$

With the aid of the foregoing results, it is easy to demonstrate the validity of the relation

$$\min_{x} \left[\int_0^T ((x', x') + 2g(x)) \, dt \right] = \max_{y} \left[-2(c, y(0)) + \int_0^T [h(y') - (y, y)] \, dt \right] \tag{9.9.1}$$

under the assumption that $g(x)$ is strictly convex. Here $x(0) = c$, $y(T) = 0$, and x and y are restrained to the appropriate function classes. The function h is the multidimensional Legendre-Fenchel transform.

9.10.   Alternate Approach

Let us next describe an alternate approach to obtaining upper and lower bounds. Consider once again the functional

$$J(u) = \int_0^T [u'^2 + 2g(u)] \, dt, \tag{9.10.1}$$

with the continuing assumption that $g(u)$ is strictly convex and that $u(0) = c$. Write

$$g(u) = \max_v \, [g(v) + (u - v)g'(v)]. \tag{9.10.2}$$

Then

$$g(u) \geq g(v) + (u - v)g'(v) \tag{9.10.3}$$

for any particular v with equality only for $v = u$. Hence, for any function $v(t)$ for which the integral exists,

$$J(u) \geq \int_0^T [u'^2 + 2ug'(v) + 2g(v) - 2g'(v)v] \, dt. \tag{9.10.4}$$

Consequently, as before we have

$$\min_u J(u) \geq \min_u \left[\int_0^T [u'^2 + 2ug'(v) + 2g(v) - 2vg(v) \,] dt \right], \tag{9.10.5}$$

and thus,

$$\min_u J(u) \geq \max_v \min_u \left[\int_0^T [u'^2 + 2ug'(v) + 2g(v) - 2vg'(v)] \, dt \right]. \tag{9.10.6}$$

For this result to be useful we must determine the minimum over u. Fortunately, this is readily done. The Euler equation associated with the minimization of

$$H(u) = \int_0^T [u'^2 + 2ug'(v)] \, dt \tag{9.10.7}$$

is

$$u'' - g'(v) = 0, \qquad u(0) = c, \qquad u'(T) = 0. \tag{9.10.8}$$

This two-point boundary-value problem is readily resolved,

$$u = c - \int_0^t \left[\int_{t_1}^T g'(v) \, dt_2 \right] dt_1 = \phi(v). \tag{9.10.9}$$

Thus, we obtain the lower bound

$$\min_u J(u) \geq \max_v \left[\varphi(v) + \int_0^T [2g(v) - 2vg'(v)] \, dt \right]. \qquad (9.10.10)$$

It is not difficult to show that the maximum is attained on the right side and that equality holds. We leave the details as a series of exercises.

EXERCISE

1. Obtain the corresponding result for the multidimensional case.

9.11 Duality

Let us now explain our continued use of the term "duality." Our fundamental tool in the scalar case has been the relation

$$g(u) = \max_v [g(v) + (u - v)g'(v)\} \qquad (9.11.1)$$

for a convex function g. Interpreting this geometrically (Figure 9.1), we see that it is equivalent to the statement that the curve lies completely above its tangent with contact only at the point of tangency. In other words, a convex curve is an envelope of its tangents.

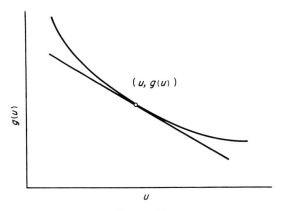

Figure 9.1

The fact that a curve may be considered as either a locus of points or an envelope of tangents, depending upon which is more convenient at the moment, is the substance of the duality underlying Euclidean geometry. This is the basis of the classical Legendre transformation, an analytic device used by Friedrichs to obtain upper and lower bounds in variational problems. We have followed the preceding path rather than that of Friedrichs, based upon a specific transformation, to indicate the flexibility of this approach, based upon the concept of convexity.

9.12.   Upper and Lower Bounds for Partial Differential Equations

The theory of dynamic programming associates a functional equation with a multistage decision process, or a process which can be interpreted to be a multistage decision process. The optimal policy is determined by the solution of the functional equation, and conversely this policy determines the solution. With the aid of this equivalence we possess simple and powerful ways of analyzing the structure of solutions of certain classes of ordinary and partial differential equations. Furthermore, with the aid of approximation in policy space we have an easy way of obtaining bounds on the solutions—upper bounds for minimization processes and lower bounds for maximization processes.

The duality techniques developed in the preceding pages enable us to associate a maximization process with a minimization process, thereby permitting us to derive both upper and lower bounds. Let us illustrate these ideas in connection with the equation

$$f_T = 2g(c) - f_c^2/4, \qquad f(c, 0) = 0, \tag{9.12.1}$$

associated with the minimization of

$$J(u) = \int_0^T (u'^2 + 2g(u))\, dt, \tag{9.12.2}$$

subject to $u(0) = c$, and the corresponding equation where $f(c, 0) = h(c)$, associated with the minimization of the functional

$$J_1(u) = \int_0^T (u'^2 + 2g(u))\, dt + h(u(T)), \tag{9.12.3}$$

again with $u(0) = c$. In what follows we will assume that both g and h are convex and that, in some fashion, we have established the fact that $f(c, T) = \min_u J(u)$ satisfies the partial differential equation for all $T \geq 0$ and all c. If not, then we can obtain only local bounds.

9.13.   Upper Bounds for $f(c, T)$

We can obtain upper bounds for the function $f(c, T)$ in two ways. To begin with, we can use the direct representation

$$f(c, T) = \min_u J(u) \tag{9.13.1}$$

to conclude that

$$f(c, T) \le J(u_1) \tag{9.13.2}$$

for any admissible function u_1, that is to say for any function u_1 such that $u_1(0) = 0$, $u_1 \in L^2(0, T)$.

Secondly, we can use approximation in policy space. Taking (9.12.1) in the form

$$f_T = \min_v [v^2 + 2g(c) + vf_c], \tag{9.13.3}$$

the form furnished by dynamic programming, we can use any policy function $v_1(c, T)$ to bound f. Namely, let $f_1(c, T)$ be determined as the solution of

$$(f_1)_T = v_1{}^2 + 2g(c) + v_1(f_1)_c, \qquad f_1(c, 0) = 0. \tag{9.13.4}$$

Then the monotonicity of the linear partial differential operator, already remarked in Chapter 6, yields the relation

$$f \le f_1 \tag{9.13.5}$$

for all c and $T > 0$. This is to be expected since $f(c, T)$ is the minimum value.

EXERCISE

1.   Write $f(c, T; g, h) = \min_u [\int_0^T (u'^2 + 2g(u))\, dt + h(u(T))]$ for g and h convex. From the fact that $g_1 \ge g_2$, $h_1 \ge h_2$ implies $f(c, T; g_1, h_1) \ge f(c, T; g_2, h_2)$, obtain a corresponding result for the associated partial differential equation.

9.14.   Lower Bounds

There are as many ways of obtaining lower bounds as there are of formulating an associated maximization process. One way, presented in Section 9.2, uses the convexity of u'^2 as a function of u'; another way, presented in Section 9.9, uses the convexity of $g(u)$. Other methods certainly exist; see Section 9.15. Which method is used in any particular case depends upon questions of analytic and computational convenience.

EXERCISE

1.   Let $f(c, T)$ be defined as in (9.12.1). For small T, obtain upper and lower estimates by using straight line approximations in the original and associated variational process.

9.15. Perturbation Technique

Consider the case where

$$g(u) = u^2 + u^4, \tag{9.15.1}$$

and $|c| \ll 1$, so that u^4 may be regarded as a perturbation term. Write

$$u^4 = \max_{v} [v^4 + 4v^3(u - v)] = \max_{v} [4uv^3 - 3v^4]. \tag{9.15.2}$$

Then

$$J(u) = \int_0^T (u'^2 + g(u)) \, dt = \int_0^T (u'^2 + u^2 + u^4) \, dt$$

$$\geq \int_0^T (u'^2 + u^2 + 4uv^3 - 3v^4) \, dt, \tag{9.15.3}$$

for all $v \in L^4(0, T)$. Hence,

$$\min_u J(u) \geq \min_u \left[\int_0^T (u'^2 + u^2 + 4uv^3) \, dt \right] - 3 \int_0^T v^4 \, dt$$

$$\geq \max_v \left\{ \left[\min_u [\cdots] \right] - 3 \int_0^T v^4 \, dt \right\}. \tag{9.15.4}$$

The minimization with respect to u can readily be carried out. Write

$$K_1(v) = \min_u [\cdots].$$

Then (9.15.4) yields

$$\min_u J(u) \geq \max_v \left[K_1(v) - 3 \int_0^T v^4 \, dt \right]. \tag{9.15.5}$$

This procedure is considerably simpler to apply than the method indicated in Section 9.7, if we are interested solely in a correction term. This is an illustration of the flexibility of the procedure based upon convexity rather than a fixed transformation.

EXERCISES

1. Show that equality holds in (9.15.5).
2. Extend the preceding to the case where $g(u) = u^2 + g_1(u)$ with $g_1(u)$ convex.
3. For the case above, use various approximations to obtain $f(c, T)$ to terms which are $0(|c|^4)$.

9.16.   Convexity of h

Let us next consider the more general case where

$$f_T = 2g(c) - f_c^2/4, \qquad f(c, 0) = h(c), \tag{9.16.1}$$

with the representation

$$f = \min_u \left[\int_0^T (u'^2 + 2g(u))\, dt + h(u(T)) \right], \tag{9.16.2}$$

$u(0) = c$.

We can proceed, as in Sections 9.2 or 9.9, using the quasi-linearizations of both g and h, that is to say, the representation of (9.11.1). We leave the details for the exercises.

EXERCISES

1. Construct the functional $K_2(v)$ defined by

$$K_2(v) = \min_u \left[\int_0^T (2g(u) - uv')\, dt + h(u(T)) \right].$$

2. Construct the functional $K_2(v, b)$ defined by

$$K_2(v, b) = \min_u \left[\int_0^T (2g(u) - uv'\, dt + h'(b)u(T)) \right].$$

3. Prove that equality holds for the corresponding maximization processes.

9.17. $2g(c) = c^2$, h Convex

The results are particularly elegant when $2g(c) = c^2$ and h is convex. The partial differential equation is

$$f_T = c^2 - f_c^2/4, \qquad f(c, 0) = h(c), \tag{9.17.1}$$

with

$$f(c, T) = \min_u \left[\int_0^T (u'^2 + u^2)\, dt + h(u(T)) \right]. \tag{9.17.2}$$

Proceeding as above, we have

$$\min_u J(u) = \max_b \left[\min_u \left[\int_0^T (u'^2 + u^2)\, dt + u(T)h'(b) \right] + h(b) - bh'(b) \right]$$

$$\tag{9.17.3}$$

The minimization with respect to u may be carried out readily, yielding the result

$$\min_u J(u) = \max_b [q_2(c, h'(b)) + h(b) - bh'(b)], \tag{9.17.4}$$

where q_2 is quadratic in c and $h'(b)$ with coefficients depending on T.

EXERCISE

1. Determine q_2 and establish (9.17.4).

9.18. The Maximum Transform

Let us now consider a transform which bears the same relation to the Legendre-Fenchel transform as the Laplace transform bears to the Fourier transform. Write

$$w(s) = M(u) = \max_{t \geq 0} [u(t) - ts], \tag{9.18.1}$$

defined for $s \geq 0$. As in the case of the Laplace transform, there may be an infinity at $s = 0$.

Our first aim is to show that $M(u)$ possesses a fundamental functional property completely analogous to the unraveling property of the Laplace transform for convolutions. Introduce the convolution-type sum

$$u * v = \max_{0 \leq s \leq t} [u(s) + v(t - s)], \qquad t \geq 0. \tag{9.18.2}$$

Then

$$M(u * v) = M(u) + M(v), \tag{9.18.3}$$

provided that $M(u)$ and $M(v)$ exist.

The proof follows a similar proof for the Laplace transform. Let us present the formal details. Consider the region $0 \leq s \leq t \leq R$ and calculate the maximum over the region in the following two ways:

$$\max_{0 \leq t \leq R} \max_{0 \leq s \leq t} = \max_{0 \leq s \leq R} \max_{s \leq t \leq R}. \tag{9.18.4}$$

Hence,

$$M(u*v) = \lim_{R \to \infty} \max_{0 \leq t \leq R} \left[\max_{0 \leq s \leq t} (u(s) + v(t - s)) - tz \right]$$

$$= \lim_{R \to \infty} \max_{0 \leq s \leq R} \left[\max_{s \leq t \leq R} \{u(s) + v(t - s) - tz\} \right]$$

$$= \lim_{R \to \infty} \max_{0 \leq s \leq R} \left[\max_{0 \leq w \leq R - s} \{u(s) + v(w) - (s + w)z\} \right]. \tag{9.18.5}$$

On the right we have

$$\max_{0 \leq s} \max_{0 \leq w} [u(s) + v(w) - sz - wz] = M(u) + M(v). \qquad (9.18.6)$$

The inversion of the maximum transform introduces no difficulties as long as we assume that $u(t)$ is concave for $t \geq 0$ and that there is a unique minimizing value $t(s)$ for each $s \geq 0$. Then it is immediate that

$$u(t) = \min_{s \geq 0} [M(u) + ts]. \qquad (9.18.7)$$

In many applications, particularly of the type discussed below, u is not concave. This introduces some delicate problems which we shall not discuss here.

EXERCISE

1.   Under what conditions is (9.18.6) a consequence of (9.18.5)?

9.19.   Application to Allocation Processes

Consider the problem of maximizing the function

$$R_N = \sum_{i=1}^{N} g_i(t_i) \qquad (9.19.1)$$

over the region D determined by

(a) $\displaystyle\sum_{i=1}^{N} t_i = c,$

(b) $t_i \geq 0,$ $\qquad\qquad\qquad\qquad\qquad\qquad\qquad\qquad (9.19.2)$

with $c > 0$. Write

$$f_N(c) = \max_{D} \left[\sum_{i=1}^{N} g_i(t_i) \right]. \qquad (9.19.3)$$

Then

$$f_1(c) = g_1(c),$$
$$f_N(c) = \max_{0 \leq t_N \leq c} [g_N(t_N) + f_{N-1}(c - t_N)], \qquad N \geq 2. \qquad (9.19.4)$$

Assuming the legitimacy of (9.18.3), it follows that

$$M(f_N) = M(g_N) + M(f_{N-1}) = \sum_{i=1}^{N} M(g_i).$$

We suppose that the $g_i(t)$ are typical utility functions, i.e., concave for $t \gg 1$, so that $M(g_i)$ exists for $s \geq 0$. The determination of f_N from (9.19.4) raises some interesting questions; see the references at the end of this chapter.

EXERCISES

1. Determine $f_N(c)$ for the case where $R_N = \sum_{i=1}^{N} a_i t_i^{b_i}, 0 < b_i < 1, a_i > 0$.
2. Determine $f_N(c)$ for the case where $R_N = \sum_{i=1}^{N} a_i t_i, a_i > 0$.
3. What modifications are necessary if we consider the cases where $\sum_{i=1}^{N} t_i \leq c$?
4. Consider the case where $\sum_{i=1}^{N} t_i \leq c_1, \sum_{i=1}^{N} b_i t_i \leq c_2$ by introducing a Lagrange multiplier and the new criterion function $R_N(\lambda) = \sum_{i=1}^{N}[g_i(t_i) - \lambda b_i t_i]$.
5. Obtain a corresponding result for the multidimensional allocation problem max $\sum_{i=1}^{N} q_i(t_i, s_i), \sum_i t_i = c_1, \sum_i s_i = c_2, t_i, s_i \geq 0$.
6. Can one use a limiting procedure to apply the maximum transform to the determination of the maximum of $J(u) = \int_0^T g(u)\, dt, \int_0^T u\, dt = c, 0 \leq u \leq k$?

9.20.   Multistage Allocation

Consider an N-stage allocation process in which at the first stage an M-dimensional vector x is divided into two nonnegative vectors y and $x - y$. As a result of this subdivision, we obtain an immediate scalar return $g(y)$, while the vector $x - y$ is transformed into $B(x - y)$ where B is a diagonal matrix with positive elements,

$$
B = \begin{pmatrix}
b_1 & & & & \\
& b_2 & & 0 & \\
& & & & \\
& 0 & & \ddots & \\
& & & & b_M
\end{pmatrix}.
\qquad (9.20.1)
$$

The process is then repeated for $N - 1$ additional stages. The problem is to determine the subdivisions at each stage so as to maximize the total N-stage return. If we write

$$f_N(x) = \text{the total return obtained from an } N\text{-stage return,}$$
$$\text{starting with } x \text{ and using an optimal policy,} \qquad (9.20.2)$$

we obtain, in the usual fashion, the recurrence relation

$$f_N(x) = \max_{0 \le y \le x} [g(y) + \lambda f_{N-1}(B(x - y))], \qquad N \ge 2,$$

$$f_1(x) = \max_{0 \le y \le x} [g(y)], \tag{9.20.3}$$

where λ is a "discount factor." $0 \le \lambda \le 1$. From this we obtain the relation

$$\phi_N(z) = h(z) + \phi_{n-1}(B^{-1}z/\lambda), \tag{9.20.4}$$

where $\phi_N = M(f_N)$, $h = M(g)$. If $f_1(x) = g(x)$, as is usually the case, we have

$$\phi_N(z) = \sum_{k=0}^{N-1} \lambda^{k-1} h(B^{-k}z/\lambda^k), \qquad N \ge 1. \tag{9.20.5}$$

If g is concave, then f_N is concave and ϕ_N is convex, thereby enabling us to employ the inversion formula given above to obtain $f_N(x)$ for any particular value of x. This procedure enables us to largely circumvent the dimensionality difficulties discussed in Chapter 3.

EXERCISE

1. Can the foregoing be extended to cover more general classes of matrices?

9.21. General Maximum Convolution

Consider the functional equation

$$\max_{x_1 + x_2 = x} [g(x_1, a) + g(x_2, b)] = g(x, h(a, b)), \tag{9.21.1}$$

where x is a vector of degree M and a of degree K. The maximum is taken over nonnegative quantities. If we can find a function $g(x, a)$ which satisfies this relation, we possess a simple way of solving the problem of maximizing

$$\sum_{i=1}^{N} g(x_i, a_i), \tag{9.21.2}$$

subject to $\sum_{i=1}^{N} x_i = x$, $x_i \ge 0$.

We can then obtain an approximate solution to the general allocation problem of maximizing $\sum_{i=1}^{N} h_i(x_i)$ over the same region by approximating to each $h_i(x)$ by a function of the form $g(x, a_i)$.

One way to study this functional equation is by way of the maximum transform. Applying it to both sides of (9.21.1), we have

$$k(y, a) + k(y, b) = k(y, h(a, b)), \tag{9.21.3}$$

where $k(y, a) = M(g(x, a))$. One solution of this equation is

$$k(y, a) = (\phi(y), a), \qquad h(a, b) = a + b. \qquad (9.21.4)$$

Other forms can be obtained by a change of variable, $a \to f(a)$. Under various assumptions we can show that all solutions are of this form. In any case, once $k(y, a)$ has been chosen, we can determine $g(x, a)$ by means of the formula

$$g(x, a) = \min_{y \geq 0} \, [k(y, a) + (x, y)]. \qquad (9.21.5)$$

EXERCISES

1. If y and a are of the same dimension show that $h(a, b) = \psi^{-1}(\psi(a) + \psi(b))$, $k(y, a) = f(y)\psi(a)$ is a solution of (9.21.3) for any ψ and f.
2. Consider the scalar case. Under what conditions on k and h are all solutions of this form? (*Hint:* Allow partial derivatives.)
3. What is the relation for the maximum of the function in (9.21.3)?

Miscellaneous Exercises

1. Consider the scalar Riccati equation $u' = a(t) - u^2$, $u(0) = c$, which may be written $u' = \min_v \, [a(t) + v^2 - 2uv]$, $u(0) = c$, and the associated linear differential equation $w' = a(t) + v^2 - 2wv$, $w(0) = c$. Show that $u = \min_v w(v, t)$.
2. Show that for any two real symmetric matrices we have $R^2 \geq SR + RS - S^2$.
3. Consider the matrix Riccati equation $R' = A - R^2$, $R(0) = I$, $A > 0$, and the associated linear equation $W' = A + S^2 - (SW + WS)$, $W(0) = I$. Show that $R \leq W(S, T)$ with equality only for $S = R$.
4. To obtain a lower bound write $R = Z^{-1}$, obtaining the equation $Z' = I - ZAZ$ and consider the associated linear equation

$$W' = (I + SAS) - (SAZ + ZAS), \, Z(0) = I.$$

(See

M. Aoki, "Note on Aggregation and Bounds for the solution of the Matrix Riccati Equation," *J. Math. Anal. Appl.* **21**, 1968, pp. 377–383.
R. Bellman, "Upper and Lower Bounds for the Solution of the Matrix Riccati Equation," *J. Math. Anal. Appl.* , **17**, 1967, pp. 373–379.
N. H. McClamroch, "Duality and Bounds for the Matrix Riccati Equation," *J. Math. Anal. Appl.*, **25**, 1969.

For some applications of the corresponding results for the scalar case to mathematical physics, see

F. Calogero, *Variable Phase Approach to Potential Scattering*, Academic Press, New York, 1967.

5. Consider the problem of minimizing $x_1(T)^2 + x_2(T)^2$ over $|f_1(t)|$, $|f_2(t)| \le 1$, $0 \le t \le T$, where $x_1' = a_{11}x_1 + a_{12}x_2 + f_1$, $x_1(0) = c_1$, $x_2' = a_{12}x_1 + a_{22}x_2 + f_2$, $x_2(0) = c_2$, an example of a "bang-bang" control process. Write

$$J(f_1, f_2) = [x_1(T)^2 + x_2(T)^2]^{1/2} = \max_b [b_1 x_1(T) + b_2 x_2(T)]$$

where the maximum is taken over $(b_1{}^2 + b_2{}^2)^{1/2} \le 1$, and

$$x_1(T) = a_1 = \int_0^T [k_{11}(T, s)f_1(s) + k_{12}(T, s)f_2(s)] \, ds,$$

$$x_2(T) = a_2 + \int_0^T [k_{21}(T, s)f_1(s) + k_{22}(T, s)f_2(s)] \, ds.$$

Show how to obtain lower bounds in this fashion (See

R. Bellman, "An Approximate Procedure in Control Theory Based on Quasilinearization," *Bull. Inst. Pol. Iasi*, **10**(14), 1964, pp. 5–10, and the monograph by Bellman *et al.* cited in 9.1 of the Bibliography and Comments.)

6. Consider the functional $J(u, v, w) = \int_0^T g(u, v, w) \, dt$, where $u' = h(u, v, w)$, $u(0) = c$. Write $f(c, T) = \min_v \max_w$. Under what conditions is $\min_v \max_w = \max_w \min_v$? If this condition holds, how can it be used to obtain upper and lower bounds for control processes?

7. Consider the special case where $g(u, v, w) = u^2 + k_1 v^2 - k_2 w^2$, $h(u, v, w) = au + v + w$.

8. Consider the nonlinear partial differential equation $f_T = \min_v \max_w [g(c, v, w) + (ac + v + w)f_c]$, $f(c, 0) = 0$. Write $\min_v [g(c, v, w) + vf_c = \phi(c, w, f_c)$; $\max_w [g(c, v, w) + wf_c] = (c, v, f_c)$. What connections are there between the solutions of $f_T = \max_w \phi(c, w, f_c)$ and $k_T = \min_v \phi(c, v, k_c)$, $f(c, 0) = k(c, 0) = 0$? (*Hint:* Consider the case where g is convex in v, concave in w.)

9. Consider the equation $v'' = t^{-1/2}v^{3/2}$, $v(0) = 1$, $v(\infty) = 0$. Show that this is the variational equation associated with the minimization of $J(u) = \int_0^\infty (u'^2 + 4t^{-1/2}u^{5/2}/5) \, dt$, that $-v'(0) = B(v) = \int_0^\infty (v'^2 + t^{-1/2}v^{5/2}) \, dt$, and that $6B(v) = 7J(v)$.

10. Show that we can obtain a maximization problem by considering $I(v) = -2v(0) - \int_0^\infty (v^2 + (6t^{1/3}/5)(v')^{5/3}) \, dt$.

11. Using trial functions of the form $u(t) = (1 + t^{1/a})^{-b}$, show that we can obtain bounds $1.5865 \leq B_0 \leq 1.5883$.

(T. Ikebe and T. Kato, "Application of Variational Method to the Thomas–Fermi Equation," *J. Phys. Soc. Japan*, **12**, 1957, pp. 201–203).

BIBLIOGRAPHY AND COMMENTS

9.1. The material in this chapter is intimately connected with the Legendre transformation, contact transformations, and the Lie theory of continuous groups. The modern approach to duality is by way of the theory of convex functions. See

E. F. Beckenbach and R. Bellman, *Inequalities*, Springer, Berlin, 1965.

S. MacLane, "Hamiltonian Mechanics and Geometry," *Amer. Math. Monthly*, **77**, 1970, pp. 570–585.

J. J. Moreau, "Convexity and Duality," *Functional Analysis and Optimization*, Academic Press, New York, 1966, pp. 145–169.

R. T. Rockafellar, *Convexity and Duality*, Princeton Univ. Press, Princeton, N. J., 1968.

L. C. Young, *Lectures on the Calculus of Variations and Optimal Control Theory*, Saunders, Philadelphia, 1968.

The equality of the minimum of the original process and the maximum of the dual process may also be derived from the fundamental min-max theorem of the theory of games, and conversely. See the foregoing references and

R. Bellman, I. Glicksberg, and O. Gross, "Some Nonclassical Problems in the Calculus of Variations," *Proc. Amer. Math. Soc.*, **7**, 1956, pp. 87–94.

S. Dreyfus and M. Freimer, *A New Approach to the Duality Theory of Mathematical Programming*, The RAND Corp. P-2334, 1961.

S. Karlin, *Duality in Dynamic Programming*, The RAND Corp. RM-971, 1952.

9.2. We have followed the path in the text to maintain as elementary a level as possible and to emphasize the versatility and flexibility of the method. The first use of the Legendre transformation to obtain a dual problem is due to Friedrichs. See

K. O. Friedrichs, "Ein Verfahren der Variationsrechung," *Nachr. Ges. Wiss. Göttingen* 1929, pp. 13–20.

The procedure we follow here was given in

R. Bellman, "Quasilinearization and Upper and Lower Bounds for Variational Problems," *Quart. Appl. Math.*, **19**, 1962, pp. 349–350.

See also

E. Kreindler, "Reciprocal Optimal Control Problems," *J. Math. Appl.*, **14**, 1966, pp. 141–152.

B. Mond and M. A. Hanson, "Duality for Variatonal Problems," *J. Math. Anal. Appl.*, **18**, 1967, pp. 355–364.

J. D. Pearson, "Reciprocity and Duality in Control Programming Problems," *J. Math. Anal. Appl.*, **10**, 1965, pp. 385–408.

G. C. Pomraming, "Reciprocal and Canonical Forms of Variational Problems Involving Linear Operators," *J. Math. Phys.*, **XLVII**, 1968.

L. B. Rall, "On Complementary Variational Principles," *J. Math. Anal. Appl.*, **14**, 1966, pp. 174–184.

R. Wilson, *The Duality Theory of Optimal Control*, to appear.
9.5. For some numerical applications, see

J. B. Yasinsky and S. Kaplan, "On the Use of Dual Variational Principles for the Estimation of Error in Approximate Solutions of Diffusion Problems," *Nuc. Sci. Eng.* **31**, 1968, pp. 80–90.
9.6. See, for example,

W. Fenchel, "On Conjugate Convex Functions," *Can. J. Math.*, **1**, 1949, pp. 73–77.
R. T. Rockafellar, "Conjugates and Legendre Transforms of Convex Functions," *Can. J. Math.*, **19**, 1967, pp. 200–205.
R. T. Rockafellar, "Duality and Stability in Extremum Problems Involving Convex Functions," *Pacific J. Math.*, **21**, 1967, pp. 167–187.
9.10. See the paper by Bellman cited in Section 9.2.
9.11. For other uses of convexity and the pole-polar relation, see the book by Beckenbach and Bellman cited in 9.1.
9.12. See

R. Bellman, "Functional Equations in the Theory of Dynamic Programming—XV: Layered Functionals and Partial Differential Equations," *J. Math. Anal. Appl.*, to appear.

R. Bellman and W. Karush, "On a New Functional Transform in Analysis: The Maximum Transform," *Bull. Amer. Math. Soc.*, **67**, 1961, pp. 501–503.
R. Bellman and W. Karush, "On the Maximum Transform and Semigroups of Transformations," *Bull. Amer. Math. Soc.*, **68**, 1962, pp. 516–518.
R. Bellman and W. Karush, "Functional Equations in the Theory of Dynamic Programming—XII: An Applicaion of the Maximum Transform," *J. Math. Anal. Appl.*, **6**, 1963, pp. 155–157.
R. Bellman and W. Karush, "On the Maximum Transform," *J. Math. Anal. Appl.*, **6**, 1963, pp. 67–74.
D. Chazan, "Profit Functions and Optimal Control: An Alternate Description of the Control Problem," *J. Math. Anal. Appl.*, **21**, 1968, pp. 169–205.
J. J. Moreau, "Fonctions convexes duales et points proximaux dans un espace Hilbertien," *C. R. Acad. Sci.*, Paris, **255**, 1962, pp. 2897–2899.
J. J. Moreau, "Analyse fonctionelle. Inf-convolution des fonctions numeriques sur un espace vectoriel," *C. R. Acad. Sci.*, Paris, **256**, pp. 5047–5049.
J. J Moreau, "Propriétes des applications 'prox'", *C. R. Acad. Sci.*, Paris, **256**, 1963, pp. 1069–1071.
P. Whittle, "The Deterministic Stochastic Transition in Control Processes and the Use of Maximum and Integral Transforms," in *Proc. Fifth Berkeley Symp. Math. Stat. Prob.*, Volume II, University of California Press, Berkeley, 1967.

For other discussion of obtaining lower bounds, see

A. W. J. Stoddart, "Estimation of Optimality for Multidimensional Control Systems," *J. Optimization Theory and Appl.*, **3**, 1969, pp. 385–391.

10

ABSTRACT CONTROL PROCESSES AND ROUTING

10.1. Introduction

In this chapter we wish to survey briefly an interesting class of problems with wide ramifications. On one hand, they can be viewed as abstractions and natural generalizations of the control processes so far encountered. On the other hand, they can be considered as particularizations which enable us to complete the foregoing investigations in some essential points, especially in connection with the computational treatment of processes of long duration. Furthermore, they provide us with an entry into the rugged terrain of scheduling processes and into the study of some other types of decision processes of combinatorial type, such as pattern recognition and numerous other problems arising in the field of artificial intelligence. Finally, in our continuing quest for feasible computational procedures, they furnish considerable motivation for the creation of sophisticated decomposition techniques based upon topological considerations. Let us also mention that questions of this nature occur more and more frequently in connection with the execution of complex computer problems.

In abstract terms, we are interested in a discrete control process of the following type: Let p be a generic element of a finite set S and $T(p, q)$ be a family of transformations with the property that $T(p, q) \in S$ whenever $p \in S$ and $q \in D$, the decision space, again taken to be discrete. We wish to determine a sequence of decisions, $q_1, q_2, \ldots$, which transform p, the initial state, into p_N, a specified state, in an optimal fashion.

We are really interested in questions of feasibility. However, in order to handle this imprecise concept we narrow our sights and consider optimization.

A problem of great contemporary interest, the "routing problem," is a particular case of the foregoing. We will use it as our leitmotif.

10.2. The Routing Problem

Consider a set of N points, numbered 1, 2, ..., N, with N the terminal point, as shown in Figure 10.1.

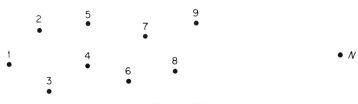

Figure 10.1

We suppose that there exists a direct link between any two points i and j which requires a time t_{ij} to traverse, with $t_{ii} = 0$, $i = 1, 2, ..., N$. In all that follows we take $t_{ij} \geq 0$. What path do we pursue, starting at 1, passing through some subsets of the points 2, 3, ..., $N - 1$, and ending at N, which requires minimum time?

Two possible paths are shown in Figures 10.2 and 10.3. The first goes directly to N; the second goes through every point before reaching N.

Figure 10.2

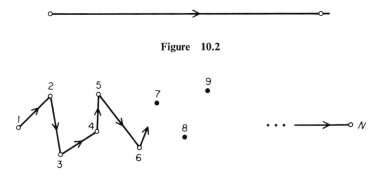

Figure 10.3

Analytically, we are required to minimize the expression

$$T(i, i_2, \ldots, i_k) = t_{1i_1} + t_{i_1 i_2} + \cdots + t_{i_k N}, \qquad (10.2.1)$$

where $(i, i_2, \ldots, i_k)$ is some subset of $(2, 3, \ldots, N-1), N)$.

EXERCISES

1. Does a minimizing path ever contain a loop, i.e., pass through the same point twice?
2. How many different admissible paths are there?
3. Can the problem presently be solved by direct enumeration with a digital computer for $N = 100$? Assume that we can enumerate and compare paths at the rate of one per microsecond.

10.3. Dynamic Programming Approach

To treat this problem by means of dynamic programming, we imbed the original problem within the family of problems consisting of determining the optimal path from i to N, $i = 1, 2, \ldots, N-1$. Let

$$f_i = \text{minimum time to go from } i \text{ to } N, \qquad i = 1, 2, \ldots, N-1,$$
$$(10.3.1)$$

and set $f_N \equiv 0$. Then the principle of optimality yields the relation

$$f_i = \min_{j \neq i} [t_{ij} + f_j], \qquad i = 1, 2, \ldots, N-1, \qquad (10.3.2)$$

with the "boundary condition" $f_N = 0$.

This functional equation is different in form, being implicit rather than explicit, from any of those so far encountered. The unknown function occurs on both sides of the equation which means that the solution cannot be obtained by any direct iteration. Consequently, some interesting analytic questions arise:

(a) Does (10.3.2) possess a solution?
(b) Does it possess a unique solution?
(c) Provided that it does possess a unique solution, how can the equation be used to provide the optimal path?
(d) How can the solution be obtained computationally?

The question of computational feasibility is, as usual, the thorniest one and one that forces us to consider new and alternate methods. Some aspects of this question will be discussed now; others will be described in the references.

EXERCISES

1. In some cases we are interested not only in the path of shortest time, but also in the next best path and generally the k-th best path. Let $f_i(2)$ be defined as time required to go from i to N following a second best path. Obtain a functional equation similar in form connecting $f_i(2)$ and f_i.
2. Similarly, obtain a functional equation for $f_i(k)$, the time associated with the k-th best path.
3. Obtain a functional equation corresponding to (10.3.2) for the case where the time to go from i to j depends upon the "direction" in which i is entered, which is to say upon the point from which i is reached.

10.4. Upper and Lower Bounds

Prior to a demonstration of the existence and uniqueness of the solution of (10.3.2), let us show how easy it is to obtain upper and lower bounds for the solution (or solutions) of (10.3.2). An "experimental proof" of uniqueness is thus available in any particular case if we can show computationally that the upper and lower bounds coincide.

Let the sequence $\{\phi_i^{(r)}\}$, $r = 0, 1, 2, \ldots$, be defined for each i in the following fashion:

$$\phi_i^{(0)} = \min_{j \neq i} t_{ij}, \qquad i = 1, 2, \ldots, N - 1, \qquad \phi_N^{(0)} = 0,$$

$$\phi_i^{(r+1)} = \min_{j \neq i} [t_{ij} + \phi_j^{(r)}], \qquad i = 1, 2, \ldots, N - 1, \qquad \phi_N^{(r+1)} = 0.$$

$$(10.4.1)$$

Let us recall that we are assuming that $t_{ij} \geq 0$.

It is clear that $\phi_i^{(r)} \geq 0$ and thus that

$$\phi_i^{(1)} = \min_{j \neq i} [t_{ij} + \phi_j^{(0)}] \geq \min_{j \neq i} t_{ij} = \phi_i^{(0)}. \qquad (10.4.2)$$

Hence, inductively,

$$\phi_i^{(0)} \leq \phi_i^{(1)} \leq \cdots \leq \phi_i^{(r)} \leq \phi_i^{(r+1)} \leq \cdots. \qquad (10.4.3)$$

Let us now show inductively that

$$\phi_i^{(r)} \leq f_i, \qquad (10.4.4)$$

where f_i is any nonnegative solution of (10.4.2). We have

$$\phi_i^{(0)} = \min_{j \neq i} t_{ij} \leq \min_{j \neq i} [t_{ij} + f_j] = f_i, \qquad (10.4.5)$$

whence (10.4.4) follows from (10.4.1) via an induction.

Since the sequence $\{\varphi_i^{(r)}\}$, $r = 1, 2, \ldots$, is uniformly bounded and monotone increasing, we have convergence for each i.. Let

$$\phi_i = \lim_{r \to \infty} \phi_i^{(r)}. \tag{10.4.6}$$

Then (10.4) yields

$$\varphi_i \leq f_i, \tag{10.4.7}$$

and (10.4.1) shows that ϕ_i is itself a solution. Hence, ϕ_i is a lower bound for any solution.

To obtain an upper bound, let us introduce the sequence $\{\psi_i^{(r)}\}$ where

$$\psi_i^{(0)} = t_{iN}, \qquad i = 1, 2, \ldots, N,$$

$$\psi_i^{(r+1)} = \min_{j \neq i} [t_{ij} + \psi_j^{(r)}], \qquad i = 1, 2, \ldots, N. \tag{10.4.8}$$

Then

$$\psi_i^{(0)} = t_{1N} \geq \min_{j \neq i} [t_{ij} + f_j] = f_i, \tag{10.4.9}$$

and again an induction establishes

$$\psi_i^{(0)} \geq \psi_i^{(1)} \geq \cdots \geq \psi_i^{(r)} \geq f_i. \tag{10.4.10}$$

Hence, the monotone decreasing sequence $\{\psi_i^{(r)}\}$, $r = 1, 2, \ldots$, converges to a limit function ψ_i which is itself a solution. Thus, for any nonnegative solution of (10.3.2), we have

$$\phi_i \leq f_i \leq \psi_i, \qquad i = 1, 2, \ldots, N, \tag{10.4.11}$$

where ϕ_i is a "lower solution" and ψ_i an "upper solution."

EXERCISES

1. Will the paths that determine $\psi_i^{(r)}$ ever possess any loops?
2. What is a geometric significance of $\psi_i^{(r)}$, and on this basis why is (10.4.10) obvious without calculation?
3. Show that $\psi_i^{(r)}$ converges in at most $N - 2$ steps.
4. Will the paths that determine $\phi_i^{(r)}$ ever possess any loops? If so, show that at most a finite number can occur if $t_{ij} > 0$. Obtain an upper bound for the number of possible loops in terms of N and $d = \min_{i,j} t_{ij}$.
5. What is the geometric significance of $\phi_i^{(r)}$?
6. Show that $\phi_i^{(r)}$ converges in a finite number of steps. Obtain an upper bound for this number of steps in terms of N and d.
7. Does the sequence $f_i^{(n)} = \min_{j \neq i} [t_{ij} + f_j^{n-1}]$, $n \geq 1$, $f_i^{(0)} = c_i$, $i = 1, 2, \ldots, N - 1, N$, converge for any nonnegative sequence $\{c_i\}$ with $c_N = 0$?

10.5. Existence and Uniqueness

The existence of *a* solution of the functional equation

$$f_i = \min_{j \neq i} [t_{ij} + f_j], \qquad i = 1, 2, \ldots, N-1,$$
$$f_N = 0, \tag{10.5.1}$$

is immediate. There is a shortest path from i to N, since there are only a finite number of admissible paths, namely those containing no loops. The time required to traverse this path (which need not be unique) defines a function g_i, $i = 1, 2, \ldots, N-1$, with $g_N = 0$.

Since a path of minimum length must go to some other point, say k, we have

$$g_i = t_{ik} + g_k \tag{10.5.2}$$

for some k. This value of k must be a value which minimizes the right-hand side. Otherwise, we would contradict the definition of g_i. Hence,

$$g_i = \min_{k \neq i} [t_{ik} + g_k], \tag{10.5.3}$$

which is to say the g_i constitute a nonnegative solution of (10.5.1).

A proof of uniqueness requires a bit more effort, but we shall give a proof for the sake of completeness. As we shall soon see, however, we have no need of this result for the purposes we have in mind.

Let us show that there cannot be two distinct solutions of (10.5.1). In order to do this, we shall assume that g_i ($i = 1, 2, \ldots, N$), and $h_i(i = 1, 2, \ldots, N)$, are two solutions, and we shall prove that $g_i = h_i(i = 1, 2, \ldots, N)$.

We know that $g_N = 0$ and $h_N = 0$. If $g_i = h_i$ for every i, there is nothing to be proved. Hence, we start with the hypothesis that there is at least one value of i for which g_i and h_i are different. Looking at all values of i for which $|g_i - h_i|$ is different from zero (if any), we now pick out the index i for which this difference is largest. By changing the numbering of the vertices, if necessary, we can suppose that $i = 1$ gives the largest difference, and by interchanging the names of g_i and h_i for every i, if necessary, we can suppose that $g_i - h_i > 0$. We now have

$$g_1 - h_1 \geqq g_i - h_i, \qquad i = 2, 3, \ldots, N. \tag{10.5.4}$$

On the other hand, from (10.5.1) we see that

$$g_1 = \min_{j \neq 1} (t_{1j} + g_j),$$
$$h_1 = \min_{j \neq 1} (t_{1j} + h_j). \tag{10.5.5}$$

Let us suppose that a value of j giving the minimum in the second equation is 2 which we can arrange by renumbering the vertices 2 to N if necessary. Then we have

$$g_1 \leqq t_{1j} + g_j, \qquad \text{for } j = 2, 3, \ldots, N,$$
$$h_1 = t_{12} + h_2, \qquad h_1 \leqq t_{1j} + h_j \qquad \text{for } j = 3, 4, \ldots, N. \tag{10.5.6}$$

These relations lead us to

$$g_1 - h_1 \leqq (t_{12} + g_2) - h_1 = g_2 - h_2. \tag{10.5.7}$$

Combining this inequality with (10.5.4), we see that $g_1 - h_1 = g_2 - h_2$.

Now we repeat this argument. We have

$$g_2 = \min_{j \neq 2} (t_{2j} + g_j),$$
$$h_2 = \min_{j \neq 2} (t_{2j} + h_j). \tag{10.5.8}$$

The value of j giving the minimum in the second equation cannot be $j = 1$, since

$$t_{21} + h_1 = t_{21} + t_{12} + h_2 > h_2. \tag{10.5.9}$$

We can therefore suppose that it is $j = 3$ (renumbering the vertices 3 to N if necessary). Then we have

$$g_2 \leqq t_{2j} + g_j, \qquad j = 1, 2, \ldots, N$$
$$h_2 = t_{23} + h_3, \qquad h_2 \leqq t_{2j} + h_j \qquad \text{for } j = 1, 2, \ldots, N. \tag{10.5.10}$$

Hence,

$$g_1 - h_1 = g_2 - h_2 \leqq t_{23} + g_2 - h_3 = g_3 - h_3. \tag{10.5.11}$$

Using (10.5.4), we therefore see that $g_1 - h_1 = g_3 - h_3$. Thus, $g_1 - h_1 = g_2 - h_2 = g_3 - h_3$.

By continuing in this manner for $N - 1$ steps, we arrive at the continued equation $g_1 - h_1 = g_2 - h_2 = \cdots = g_i - h_i$ ($i = 1, 2, \ldots, N - 1$), thus proving that the two solutions are in fact identical.

We have previously shown that the desired minimal times comprise a solution of (10.5.1). Thus it follows from the uniqueness just proved that if we can by any method whatsoever find a solution $f_1, \ldots, f_{N-1}$ of (10.5.1) then this solution provides us with the desired minimal times.

EXERCISE

1. Suppose that not all points i and j are connected by a link. Show that we can reduce this more common case to the foregoing case, for compu-

tational purposes, by assuming that there exists a link with a large time associated whenever no link exists in the original network, and give a value for this "large time."

10.6.   Optimal Policy

Once we have established uniqueness of the solution of (10.5.1), we can use this solution to determine the optimal policy, a function $j(i)$ which tells us what point (or points) to go to from i. It may be obtained by minimizing the quantity $t_{ij} + f_j$ with respect to j.

An optimal path is then given by

$$[i, j_1, j_2, \ldots, j_k, N], \tag{10.6.1}$$

where

$$
\begin{aligned}
j_1 &= j(1), \\
j_2 &= j(j_1), \\
&\vdots \\
j_k &= j(j_{k-1}), \\
N &= j(j_k).
\end{aligned}
\tag{10.6.2}
$$

EXERCISES

1.   Determine the forms of some approximate policies.

10.7.   Approximation in Policy Space

The equation

$$
\begin{aligned}
f_i &= \min_{j \neq i} [t_{ij} + f_j], \qquad i = 1, 2, \ldots, N - 1, \\
f_N &= 0,
\end{aligned}
\tag{10.7.1}
$$

can be solved by means of successive approximations as we have indicated. If N is large, the choice of an initial approximation can be crucial in determining the time required for the calculation.

In many cases, a great deal is known about the underlying process, which means that various types of approximate policies exist. It may then be far better to approximate in policy space than in function space. This is a common situation in the study of control processes.

Let $j_0(i)$ be a policy which enables us to go from i to N and let $f_i^{(0)}$ be calculated using this policy,

$$f_i^{(0)} = t_{ij_0} + f_j^{(0)} = t_{ij_0} + \cdots, \tag{10.7.2}$$

where the dots indicate the terms obtained by iterating the relation. This furnishes an initial approximation which can be used as a starting point for the method described above.

An interesting and important question is whether we can obtain monotone approximation in policy space. Given any initial policy $j_0(i)$ we want a systematic way of obtaining a new policy which yields times no larger than the original, and smaller if the original policy is not an optimal policy. One way of doing this is the following: With $f_i^{(0)}$ determined as in (10.7.2) determine j to minimize the expression $t_{ij} + f_i^{(0)}$. Call this function $j_1(i)$. We assert the function $f_i^{(1)}$ calculated using the policy $j_1(i)$ satisfies the relation $f_i^{(1)} \leq f_i^{(0)}$.

This follows from the relations

$$\begin{aligned} f_i^{(0)} &= t_{ij_0} + f_j^{(0)} \geq t_{ij_1} + f_{j_1}^{(0)}, \\ f_i^{(1)} &= t_{ij_1} + f_j^{(1)}. \end{aligned} \tag{10.7.3}$$

Observe again how the technique of approximation in policy space is tied in with the theory of positive operators.

10.8.   Computational Feasibility

Let us now examine the computational feasibility of the methods of successive approximation given in Section 10.4. Let us consider the second method,

$$\begin{aligned} \psi_i^{(0)} &= t_{iN}, \\ \psi_i^{(r+1)} &= \min_{j \neq i} [t_{ij} + \psi_i^{(r)}], \qquad i = 1, 2, \ldots, N. \end{aligned} \tag{10.8.1}$$

To calculate the set of values $\{\psi_i^{(r)+1}\}$, $i = 1, 2, \ldots, N$, we require the previous set $\{\psi_i^{(r)}\}$ and the matrix (t_{ij}). If N is large, e.g., $N = 10^4$, storage of the matrix (t_{ij}) involving $N^2 = 10^8$ values puts an unbearable strain on rapid-access storage. Observe, however, that the situation is a good deal better than it first apppears because to calculate $\psi_i^{(r+1)}$, for a particular value of i requires only that the column

$$t_i = \begin{pmatrix} t_{i1} \\ t_{i2} \\ \vdots \\ t_{iN} \end{pmatrix} \tag{10.8.2}$$

and $\{\psi_i^{(r)}\}$, $i = 1, 2, 3, \ldots, N$, be available in rapid-access storage. Naturally, we pay a cost in time for this shifting of data back and forth from core, but this is bearable compared with the cost of not being able to treat the problem at all. If, for example, $N = 10^{10}$, the algorithm as it stands is not feasible at the present time even with this artifice.

EXERCISE

1. Is there any advantage in time or storage in picking the K first points to minimize over, $K \geq 2$?

10.9.   Storage of Algorithms

In the previous section we indicated that severe difficulties were encountered if we insisted upon the values t_{ij} being immediately accessible. Suppose that the situation is such that we can calculate t_{ij} as needed using an algorithm which requires very few instructions. It will, in general, require a certain amount of time to calculate t_{ij} as opposed to looking it up in a table. This means that once again we are trading time, which we possess, for rapid-access storage capacity on which there is an absolute upper bound.

As an example of this, consider the case where the ith point is specified by Cartesian coordinates (x_i, y_i) and where the time to traverse the distance between i and j is directly proportional to this distance. Taking this factor of proportionality to be unity, we have

$$t_{ij} = [(x_i - x_j)^2 + (y_i - y_j)^2]^{1/2}. \tag{10.9.1}$$

If N is small, we can calculate these numbers in advance; if N is large, it may be better to generate them as they are needed. This technique enables us to handle routing processes of high dimension with very little demand on rapid-access storage or core.

10.10.   Alternate Approaches

The routing problem has attracted a great deal of attention as a consequence of an almost unique combination of intrinsic simplicity of statement and widespread application. As a consequence, a number of ingenious and effective techniques now exist based on principles quite different than those expounded above.

This is very important since it means that some of these techniques may be applicable to the solution of a number of control problems using the ideas to be discussed in Section 10.13.

10.11.   "Travelling Salesman" Problem

Let us now discuss an interesting example of a combinatorial problem where the introduction of suitable state variables allows us to use the foregoing approach to the routing problem. In place of asking for a path of minimal time from 1 to N, let us now seek to determine a path starting at 1, passing through each point 2, 3, ..., N in some order, and returning to 1, which consumes the least time. It is clear why this is called the "traveling salesman" problem.

We see that at each point in the tour it is necessary to know either where we have been, or where we have not yet been. It is not sufficient, as in the previous process, to keep track only of the current position.

Let us then introduce a multistage decision process in which the state of the system at any time is denoted by

$$p = [i_1 i_2 \cdots i_{k-1} i_k] \tag{10.11.1}$$

where i_1, i_2, ..., i_{k-1} denotes points already visited starting from the origin the point 1, and where i_k is the current point; see Figure 10.4.

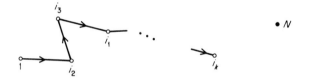

Figure   10.4

We have drawn the diagram (Figure 10.4) to emphasize the fact that the *order* in which i_1, i_2, ..., i_{k-1} appear in (10.11.1) is of no importance. Once we are at i_k, it suffices to know that we have already visited i_1, i_2, ..., i_{k-1} in some order. This turns out to be an essential feature as far as computational feasibility of the following method is concerned.

The principle of optimality yields the functional equation

$$f(i_1, i_2, \ldots, i_k) = \min_{j} \, [t_{i_k j} + f(i_1, i_2, \ldots, i_k, j)], \tag{10.11.2}$$

where

$$f(i_1, i_2, \ldots, i_k) = \text{time required to tour the remaining points}$$
$$\text{and return to 1 using an optimal policy.} \tag{10.11.3}$$

When $i_k = N$ and i_1, i_2, ..., i_{k-1} represent 2, 3, ..., $N - 1$ in some order, we have

$$f(i_1, i_2, \ldots, i_k) = t_{i_k 1}. \tag{10.11.4}$$

Computational feasibility will be briefly treated in the exercises. Here we were principally concerned with the idea of an *induced* routing process.

EXERCISES

1. What rapid-access storage will be required to solve an N-point "traveling salesman" problem, assuming that at the proper stage we change over from keeping track of where we have been to keeping track of where we still have to go?

2. What slow-access storage (core) would be required to treat the foregoing problem for the case where the points are the fifty state capitols?

3. Are there any advantages to choosing the K first points in the path, solving the remaining $(N - K)$-point traveling salesman problem and then maximizing over the N first points?

4. Devise some methods of successive approximation which would allow one to obtain reasonable estimates of the minimum time required for the case $N \gg 1$ assuming "time" is equal to distance.

5. A tourist wants to visit N points of interest, seeing no more than M on a particular day and returning to his hotel each night. What is an efficient way of doing this? (See

   R. Bellman, "An Algorithm for the Tourist Problem," *Bull. Inst. Math. Its Appl.*, to appear).

10.12. Stratification

In general, regardless of the power of the computer available in the foreseeable future, complex processes cannot be handled without making use of intrinsic structural features. It is thus the recognition and exploitation of structure that is the art of mathematics. Our aim here is to utilize this structure to reduce dimensionality difficulties, to lift the "curse of dimensionality."

As an example of the way in which the structure of the process can greatly simplify our task, consider the case where the set of points has the appearance in Figure 10.5.

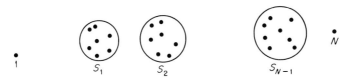

Figure 10.5

The point of the figure is to indicate the fact that starting at 1, we must go to some point in S_1, from this point to some point in S_2, and so on; from a point in S_{N-1} we must go to N. Rather than treat this in the foregoing general fashion, we introduce a sequence of functions

$$f_i(p) = \text{minimum time from the point } p \in S_i \text{ to } N, \qquad i = 1, 2, \ldots, N-1$$
$$(10.12.1)$$

Then we readily obtain the recurrence relations

$$f_i(p) = \min_{p_1 \in S_{i+1}} [t_{p, p_1} + f_{i+1}(p_1)], \qquad (10.12.2)$$

for $i = 1, 2, \ldots, N-1$, with

$$f_{N-1}(p) = t(p, N), \qquad (10.12.3)$$

with $t(p, p_1)$ the time to go from p to p_1.

If f_1 is as before, the minimum time to go from 1 to N, we have

$$f_1 = \min_{p \in S_1} [t(1, p) + f_1(p)]. \qquad (10.12.4)$$

It is clear that the demands on rapid-access storage have been significantly reduced by means of this decomposition or *stratification*, as we shall call it. In a number of important processes, time automatically yields this stratification. By this we mean that the set S_k will be the set of admissible states at time k.

When there is no variable corresponding to time, or when the dimensions of the sets S_k are themselves overwhelming, the process of stratification becomes much more complex and represents a new type of topological-algebraic problem. We shall discuss in Section 10.17 a process of this nature. As might be expected, the type of stratification possible depends critically upon the nature of the underlying control process.

10.13.   Routing and Control Processes

Let us now turn to an examination of the application of the foregoing ideas to control processes and, particularly, to their computational treatment. Consider our pet example, the problem of minimizing the functional

$$J(u) = \int_0^T (u'^2 + 2g(u)) \, dt, \qquad (10.13.1)$$

where $u(0) = c$ and g is convex. As we know, for any $T > 0$ the minimizing function exists and is specified as the unique solution of the Euler equation

$$u'' - g'(u) = 0, \qquad u(0) = c, \qquad u'(T) = 0. \qquad (10.13.2)$$

Specified yes, but is it constructively determined?

We have shown that crude successive approximation techniques will work when $T \ll 1$ and that a more sophisticated technique based upon quasilinearzation will be effective provided we can obtain a plausible initial approximation. We have also indicated the use of Bubnov-Galerkin and Rayleigh-Ritz techniques to obtain the required initial approximation. Other techniques based upon direct and sophisticated search techniques exist. Let us now present a quite different approach based upon a combination of the routing algorithm and the calculus of variations.

If $T \ll 1$, as just mentioned, we can determine the minimizing function. We thus possess efficient techniques for obtaining local solutions. Can we piece together local solutions, $T \ll 1$, in some fashion to determine global solutions, $T \gg 1$? It turns out that we can. We can use the calculus of variations to obtain these local solutions and dynamic programming to combine these local solutions and thus locate the global solution or, at least, its desired approximation. This is a nice example of the importance of combining different methods.

10.14.   Computational Procedure

Let us sketch briefly how we might proceed. We begin with a grid in the (c, T)-plane, as in Figure 10.6. Suppose that at the times $T_1, T_2, \ldots, T_k$, where $0 < T_1 < T_2 < \cdots < T_k < T$, we allow only the states $c_1, c_2, \ldots, c_r$.

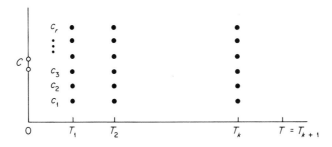

Figure   10.6

By this we mean that if the state is c_i at time T_k, it is required that the state of the system become c_j at times T_{k+1}. This is a suboptimization technique. Furthermore, we require that the path from c_i to c_j be a "geodesic" in the sense that it minizes the functional

$$J(T_k, T_{k+1}, c_i, c_j ; u) = \int_{T_k}^{T_{k+1}} u'^2 + 2g(u)) \, dt, \qquad (10.14.1)$$

with u subject to the constraints

$$u(T_k) = c_i, \qquad u(T_{k+1}) = c_j. \tag{10.14.2}$$

When $T_{k+1} = T$, there is no terminal constraint.

If $T_{k+1} - T_k$ is not too large and if $|c_i - c_j|$ is not too large, we know successive approximations, of either simple or complex type, will yield the quantity

$$t_{ij} = \min_u J(T_k, T_{k+1}, c_i, c_j; u). \tag{10.14.3}$$

We can ensure this by introducing a "nearest neighbor" constraint. We allow only those values of j for which $|c_i - c_j| \ll 1$. In other words, we are looking for a reasonably smooth trajectory.

Once the matrix of these fundamental times has been determined, we can apply the routing algorithm to determine an optimal path of the foregoing nature over $[0, T]$. This path will have, in general, the appearance of Figure 10.7.

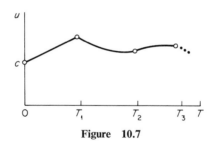

Figure   10.7

We now use this as an initial approximation for quasilinearization applied to the Euler equation over $[0, T]$, and we proceed to obtain a more accurate representation of the minimizing function.

EXERCISE

1.   How would we go about smoothing this initial approximation to obtain one with a continuous derivative at T_1, T_2, etc.? Is this necessary for the quasilinearization method?

10.15.   Feasibility

Let us take a closer look at what is required for a successful application of the foregoing method. To begin with, it is necessary to have a reasonable idea of the part of state space of interest. In some cases, there exist a priori bounds on the domain of the state of the system, either established analytically or on the basis of physical considerations, which furnish this information. Secondly, there is the question of determining the allowable steps in time and the grid in state space.

As usual, we are on the horns of a dilemma, or perhaps polylemma. If the timestep is small, we can guarantee the calculation of t_{ij}. This decrease in $|t_{i+1} - t_i|$, however, is at the expense of increasing the time of the entire calculation. Furthermore, even if the timestep is small, we have to worry about abrupt changes of state. In other words, as previously indicated, it may be wise to impose some bounds on the change of state, which would again be a restriction of either analytic or physical origin. This means that in minimizing over p_1 in the relation (10.11.2), we want to impose a restriction $p_1 \in R(p)$ where $R(p)$ consists of states near p in an appropriate norm. This is a "nearest neighbor" constraint: In short time intervals we can only go to nearby points in state space. Restrictions of this type speed up the calculation to a great extent, greatly facilitate the determination of the t_{ij}, and usually represent meaningful constraints on the type of control that is allowable.

10.16.   Perturbation Technique

In many cases, we possess an initial approximation to the minimizing function, say u_0. Sometimes, we want to obtain more and more refined approximations; sometimes we merely want to know whether the approximation is a reasonable one.

One way to proceed is as follows: We write

$$u = u_0 + w,$$

$$g(u) = g(u_0 + w) = g(u_0) + wg'(u_0) + w^2/2 \, g''(u_0) + \cdots, \quad (10.16.1)$$

and consider the minimization of the quadratic functional

$$J_1(w) = \int_0^T [(u_0' + w')^2 + 2g(u_0) + 2wg'(w_0) + w^2 g''(u_0)] \, dt. \quad (10.16.2)$$

As we have previously noted, this is equivalent to applying quasilinearization to the Euler equation. If the minimizing w is small, we have the desired improvement; if it is large, we have been warned that u_0 is not a good choice. There are many ways of treating the quadratic variational problem in (10.16.2).

An alternate approach is to draw a tube around the initial approximation and then proceed as we have to stratify state space; see Figure 10.8.

We hope that the routing algorithm yields a result of the type in Figure 10.9.

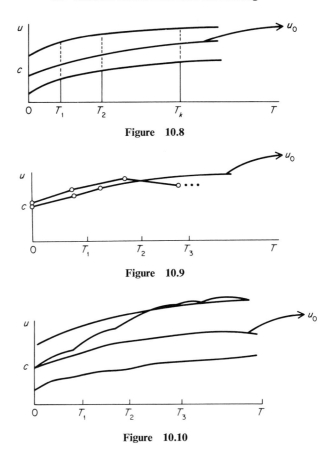

Figure   10.8

Figure   10.9

Figure   10.10

If, on the other hand, the routing algorithm yields a solution which keeps hitting the boundary of the tube, as in Figure 10.10, we can conclude that u_0 was not a good initial approximation.

Let us again emphasize that there is nothing routine about determining optimal policies in realistic control processes. If we hope to obtain optimal policies, we must enter the fray armed with an arsenal of weapons and be psychologically prepared for the use of various combinations of methods. Successful computing is an adaptive control process.

10.17.   Generalized Routing

In some cases we encounter in place of the process originally associated with Figure 10.5 a process where from a point in S_i we can go either to another point in S_i or to S_{i+1}. A transformation of the first type we call *recurrent*;

one of the second type is *transitive*. In place of (10.12.2) we obtain the recurrence relation

$$f_i(p) = \min \left[\min_{p_1 \in S_i} [t(p, p_1) + f_i(p_1)], \min_{p_2 \in S_i} [t(p, p_2) + f_{i+1}(p_2)] \right]. \quad (10.17.1)$$

In other cases, the situation is still more complex. We encounter a situation where the set of points has the form in Figure 10.11. We have a branching of states and of sets. Once again, a point may be transformed into another point in the same set or into one of the succeeding sets.

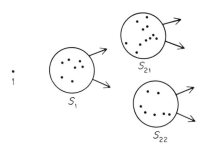

Figure 10.11

Provided that there is not too much branching, we can use the routing algorithm to determine the optimal paths. In general, the task of constructing a decomposition of this type is not an easy one.

EXERCISES

1. Obtain functional equations analogous to those of (10.17.1) corresponding to Figure 10.11.
2. As an example of a process where the branching gets out of hand, consider the problem of calculating 2^N with a minimum number of multiplications.

10.18. **Pawn-King Endings in Chess**

As an example of a situation where a suitable stratification provides us with a feasible computational approach, consider the determination of optimal play in a King-Pawn ending in chess. If we allow a King and two Pawns for each side, we can contemplate at worst about $(60)^6$ possible positions and, generally, far fewer if the Pawns are opposing and if we consider positions of interest. If we assume a "boundary condition" of perfect determination of a win, loss or draw as soon as a Pawn promotes, we have a multistage process which is almost within the rapid-access realm of current computers using a direct routing algorithm.

We can, however, bring it easily within the reach of contemporary abilities by an appropriate stratification. Observe that in King-Pawn endings there are two types of moves—those which are reversible and those which are irreversible. The reversible moves are King moves not involving a capture; the irreversible moves are those involving a capture or the move of a Pawn. These are, respectively, the recurrent and the transitive transformations described above. A detailed discussion of these matters will be found in the Bibliography.

EXERCISES

1. Write a functional equation for the determination of optimal play for the case where there is a King against a King and a Pawn.
2. Write a functional equation for the determination of optimal play for the case where there is a King against a King and a Rook. In particular, determine the play which minimizes the number of moves required to checkmate.

10.19.   Discussion

What we have been emphasizing is the fact that appropriate utilization of the intrinsic structure of a process can convert an intractable formulation into one that leads to a quick and simple calculation. The problems encountered in this type of endeavour are highly algebraic and topological. Essentially what is required is a semi-group theory of structures. Some few tentative steps have been taken in this direction, but almost everything remains to be done.

Miscellaneous Exercises

1. Let x_i be either an a or b and consider expressions of the form $x_1 x_2 \cdots x_N$. Suppose that these have the property that $(x_1 x_2 \cdots x_N)(x_1 x_2 \cdots x_N)$ reduces to an identity element. Show that the only expressions that cannot be condensed are 1, a, b, ab, ba, aba, bab.
2. Consider the same situation where $(x_1 x_2 \cdots x_N)^3 = 1$. Are there a finite or infinite number of non-equivalent expressions?
3. Let $f_N(a)$ be the length of the "word" of maximum length starting with a when at most N multiplications are allowed. Show that $f_N(a) = 1 + \max_x f_{N-1}(ax)$, $n \geq 2$, with a similar expression for $f_N(b)$.

4. Discuss the feasibility of the foregoing as a computational scheme for determining $f_N(a)$. (The foregoing is stimulated by the Burnside conjecture in group theory.)

BIBLIOGRAPHY AND COMMENTS

10.1. Detailed discussions and many additional references may be found in

R. Bellman, K. L. Cooke, and J. Lockett, *Algorithms, Graphs, and Computers*, Academic Press, New York, 1970.

See also

R. Bellman, *Dynamic Programming and Problem-Solving*, TR-69-2, Univ. Southern California, 1969.

R. Bellman and K. L. Cooke, "The Königsberg Bridges Problem Generalized," *J. Math. Anal. Appl.*, **25**, 1969, pp. 1–7.

10.2. For some economic applications, see

R. Radner, "Paths of Economic Growth that are Optimal with Regard Only to Final States: A 'Turnpike Theorem'," *Rev. Econ. Stud.*, **28**, 1961, pp. 98–104.

10.3. The method was first presented in

R. Bellman, "On a Routing Problem," *Quart. Appl. Math.*, **16**, 1958, pp. 87–90.

See also

R. Bellman and R. Kalaba, "On k-th Best Policies," *J. Soc. Indus. Appl. Math.*, **8**, 1960, pp. 582–588.

10.7. See the book

R. Howard, *Dynamic Programming and Markovian Decision Processes*, M.I.T. Press, Cambridge, Mass., 1965.

where the "policy improvement" method is disscussed in detail, and

G. J. Minty, "Monotone Networks," *Proc. Roy. Soc. London*, Ser. A, **257**, 1960, pp. 194–212.

10.10. A careful and thorough analysis and comparison of a number of different methods now available may be found in the paper,

S. Dreyfus, *An Appraisal of Some Shortest Path Algorithms*, The RAND Corp., RM-5433-PR, October, 1967.

J. Monavek, "A Note upon Minimal Path Problems," *JMAA*, 1970, to appear.

See also the book cited in Section 10.1.

10.11. The method was first presented in

R. Bellman, "Dynamic Programming Treatment of the Traveling Salesman Problem," *J. Assoc. Comput. Mach.*, **9**, 1962, pp. 61–63.

See also

R. Bellman, "Combinatorial Processes and Dynamic Programming," *Proc. Ninth Symp. Appl. Math.*, American Mathematical Society, 1960.

V. N. Burkov and S. E. Lovetskii, "Methods for the Solution of Extremal Problems of Combinatorial Type" (Review), *Autom. and Remote Control*, no. 11, 1968, pp. 1785–1806.

For use of these ideas in the field of pattern recognition see

K. S. Fu, *Sequential Methods in Pattern Recognition and Machine Learning*, Academic Press, New York, 1968.

When the number of points becomes large, $N \geq 50$, various approximate methods must be employed. An excellent survey of the current state of the problem is given in

R. E. Gomory, "The Travelling Salesman Problem," *Proc. IBM Sci. Comput. Symp. Combinatorial Probl.*, 1964.
    10.12.   The concept of stratification is discussed in

R. Bellman, *Dynamic Programming and Markovian Decision Processes with Particular Applications to Baseball and Chess*, *Applied Combinatorial Mathematics*, Wiley, New York, 1964, pp. 221–236.
R. Bellman, "Stratification and the Control of Large Systems with Application to Chess and Checkers," *Inform. Sci.*, **1**, 1968, pp. 7–21.

with application to the determination of optimal play in chess and checkers.
    See also

R. Larson, *State Increment Dynamic Programming*, Elsevier, New York, 1968.
E. L. Lawler, "Partitioning Methods for Discrete Optimization Problems," in *Recent Mathematical Advances in Operations Research*, (E. L. Lawler and G. Minty, eds.), Univ. of Michigan, Ann Arbor, 1964.

and the works of Aris and Nemhauser previously referenced.
    10.13.   See

J. F. Shapiro, *Shortest Route Methods for Finite State Space, Deterministic Dynamic Programming Problems*, Tech. Rep. Operations Research Center, M.I.T., Cambridge, Mass., 1967.
    10.16.   See the book

D. H. Jacobson and D. Q. Mayne, *Differential Dynamic Programming*, Elsevier, New York, 1970.

for a systematic discussion of this method.
    10.18.   See the paper in *Information Sciences* referred to in 10.12.
    10.19.   See

H. Happ, *The Theory of Network Diakoptics*, Academic Press, New York, 1970.
B. K. Harrison, "A Discussion of Some Mathematical Techniques Used in Kron's Method of Tearing," *J. Soc. Indus. Appl. Math.*, **11**, 1963.
G. Kron, "Tearing and Interconnecting as a Form of Transformation," *Quart. Appl. Math.*, **13**, 1955, pp. 147–159.
G. Kron, "A Set of Principles to Interconnect the Solution of Physical Systems," *J. Appl. Phys.*, **24**, 1953, pp. 965–980.
J. P. Roth, "An Application of Algebraic Topology, Kron's Method of Tearing," *Quart. Appl. Math.*, **17**, 1959, pp. 1–14.
W. R. Spillers and N. Hickerson, "Optimal Elimination for Spaers Symmetric Systems as a Graph Problem," *Quart. Appl. Math.*, **26**, 1968, pp. 425–432.

D. V. Stewart, "Partitioning and Tearing Systems of Equations," *J. Soc. Indust. Appl. Math.*, Numer. Anal., **2**, 1965, pp. 345–365.

For an interesting discussion of scheduling theory, see

L. R. Ford, Jr. and D. R. Fulkerson, "Constructing Maximal Dynamic Flows from Static Flows," *Oper. Res.*, **6**, 1958, pp. 419–433.

See also

B. L. Fry, *Network-Type Management Control Systems Bibliography*, The RAND Corp. RM-3047-PR, 1963.

11

REDUCTION OF DIMENSIONALITY

11.1. Introduction

In preceding chapters on some computational aspects of the calculus of variations and dynamic programming, we have emphasized the serious difficulties that can arise when either of these general theories are applied to the study of the control of systems of large dimension. These formidable obstacles are clear indications that there is a pressing need for new approaches and, indeed, a separate theory of large systems.

Two general methods will certainly be integral parts of this new theory—one is the method of successive approximations and the other is the technique of decomposition. In this chapter, we shall focus on approximations. In Chapter 10, we gave an example of what is meant by decomposition of a process.

What is novel in our use of approximation is that our goal in utilizing this traditional technique is not the usual one of obtaining a linear system. Rather, it is to arrive at a system, linear or nonlinear, which is of low enough dimension to permit easy computational treatment using any of a number of analytic approaches. What is clear is that the "dimension" of a control process is not intrinsic but is strongly dependent on the analytic techniques employed.

We shall present some results for systems governed by ordinary differential equations which illustrate these remarks. As will be seen, the results obtained are quite meager, illustrating the need for a great deal of additional research and, what is much more difficult, some new ideas.

11.2. A Terminal Control Process

Let us begin by considering the following terminal control problem. We wish to minimize the quadratic functional

$$J(x, y) = (x(T), Bx(T)) + \int_0^T (y, y) \, dt, \tag{11.2.1}$$

where x and y are connected by the linear relation

$$x' = Ax + y, \qquad x(0) = c. \tag{11.2.2}$$

We have already indicated in Chapters 3 and 5 some of the computational obstacles associated with this simple analytic problem in the case where x is of large dimension. One type of difficulty is encountered using the calculus of variations; another occurs when dynamic programming is employed. Here we wish to consider the case where the matrix B has the special form

$$B = \begin{pmatrix} \begin{pmatrix} b_{11} & \cdots & b_{1k} \\ b_{21} & \cdots & b_{2k} \\ \vdots & & \\ b_{k1} & \cdots & b_{kk} \end{pmatrix} & 0 \\ & & \\ 0 & & 0 \end{pmatrix}, \tag{11.2.3}$$

with $k \ll N$.

This is equivalent to the statement that in evaluating the terminal behavior of the system only the first k components of $x(T)$ are of significance, a situation which occurs frequently in applications. The other components, x_{k+1}, $x_{k+2}, \ldots, x_N$, play the role of "hidden variables." Although they are vital for a classical deterministic description of the system, we have no specific interest in their precise behavior at the end of the process.

A particular case of the foregoing general problem is that where we wish to minimize an expression of the form

$$J(u, v) = \sum_{i=1}^k b_i u^{(i)}(T)^2 + \int_0^T v^2 \, dt, \tag{11.2.4}$$

with u and v connected by the relation

$$u^{(N)} + a_1 u^{(N-1)} + \cdots + a_N u = v,$$
$$u^{(i)}(0) = c_1, \qquad i = 0, 1, \ldots, N - 1. \tag{11.2.5}$$

11.3.   Preliminary Transformation

The linearity of (11.2.2) is the basic factor. We begin by noting the relation

$$x(t) = e^{At}c + \int_0^t e^{A(t-s)}y(s)\, ds, \qquad (11.3.1)$$

whence

$$x(T) = e^{AT}c + \int_0^T e^{A(T-s)}y(s)\, ds. \qquad (11.3.2)$$

We presage our subsequent procedure by writing this in the form

$$x(T) = z + \int_0^T K(T-s)y(s)\, ds, \qquad (11.3.3)$$

where z is a vector parameter and K is a given matrix. This parameter z has an extremely interesting physical significance; namely it is the state of the system at time T if no control is exerted at all, i.e., if $y = 0$, $0 \le t \le T$.

The i-th component of $x(T)$ then has the form

$$x_i(T) = z_i + \int_0^T \left[\sum_{j=1}^N k_{ij}(T-s)y_j(s) \right] ds. \qquad (11.3.4)$$

Hence, if B has the special form indicated in (11.2.3), we see that $J(x, y)$ takes the form

$$\sum_{i,j=1}^k b_{ij}\left(z_i + \int_0^T [\cdots]\, ds \right)\left(z_j + \int_0^T [\cdots]\, ds \right) + \int_0^T (y, y)\, dt, \quad (11.3.5)$$

where $[\cdots]$ contains terms linear in the y_j.

11.4.   New State Variables

Starting from (11.3.5), we consider the z_i as new state variables and introduce the new criterion function

$$\phi(z, T) = \min_y \left[\sum_{i,j=1}^k b_{ij}\left(z_i + \int_0^T [\cdots]\, ds \right)\left(z_j + \int_0^T [\cdots]\, ds \right) + \int_0^T (y, y)\, dt \right],$$

$$(11.4.1)$$

defined for $T \ge 0$, $-\infty < z_i < \infty$, $i = 1, 2, \ldots, k$.

What is of basic importance in what follows is that $\phi(z, T)$ is a function of only k variables. We have thus converted the original optimization problem involving a function of N state variables into one involving k state variables. If $k \ll N$, this new problem is much easier to treat. In particular, a dynamic programming solution is now feasible in a number of cases.

11.5. Partial Differential Equation for $\phi(z, T)$

Let us deduce a partial differential equation for $\varphi(z, T)$ using familiar reasoning. Decompose the interval $[0, T]$ into $[0, \Delta]$ and $[\Delta, T]$. Write, as usual,

$$\min_{y[\Delta, T]}, \quad \min_{y[0, \Delta]} \tag{11.5.1}$$

to denote, respectively, minimization with respect to y chosen over $[\Delta, T]$ and $[0, \Delta]$. Then we have

$$\phi(z, T) = \min_{y[0, \Delta]} \min_{y[\Delta, T]} \left[\sum_{i,j=1}^{k} b_{ij} \left(z_i + \int_0^{\Delta} + \int_{\Delta}^{T} \right) \left(z_j + \int_0^{\Delta} + \int_{\Delta}^{T} \right) \right.$$

$$\left. + \int_0^{\Delta} (y, y) \, dt + \int_{\Delta}^{T} (y, y) \, dt \right]. \tag{11.5.2}$$

We can proceed formally in our derivation, secure in the knowledge that all can easily be made rigorous using the fact that ϕ is a quadratic form in z. This analytic structure will be explicitly used in the following section.

The integral from 0 to Δ is approximately

$$\int_0^{\Delta} \left(\sum_{j=1}^{N} k_{ij}(T - s) y_j(s) \right) ds = \left[\sum_{j=1}^{N} k_{ij}(T) y_j(0) \right] \Delta + o(\Delta), \tag{11.5.3}$$

while the integral over $[\Delta, T]$ can be written

$$\int_{\Delta}^{T} \left(\sum_{j=1}^{N} k_{ij}(T - s) y_j(s) \right) ds = \int_0^{T-\Delta} \left[\sum_{j=1}^{N} k_{ij}(T - \Delta - s) y_j(s + \Delta) \right] ds. \tag{11.5.4}$$

Hence, we see that T is transformed into $T - \Delta$ and z into $z + K(T) y(0)\Delta$. Thus, the principle of optimality yields the relation

$$\phi(z, T) = \min_{y(0)} [(y(0), y(0))\Delta + \phi(z + K(T)y(0)\Delta, T - \Delta) + o(\Delta)]. \tag{11.5.5}$$

Letting Δ approach zero, the result is the partial differential equation

$$\frac{\partial \phi}{\partial T} = \min_{y(0)} [(y(0), y(0)) + (K(T)y(0), \text{grad } \phi)]$$

$$= \min_{y(0)} [(y(0), y(0)) + (y(0), K(T)' \text{ grad } \phi)], \tag{11.5.6}$$

where $K(T)'$ here denotes the transpose of $K(T)$.

The minimum is achieved at

$$y(0) = \frac{-K(T)' \text{ grad } \phi}{2},$$

(11.5.7)

leading to the relation

$$\frac{\partial \phi}{\partial T} = -\left(\frac{K(T)' \text{ grad } \phi}{2}, \frac{K(T)' \text{ grad } \phi}{2} \right).$$

(11.5.8)

The initial condition is

$$\phi(z, 0) = (z, Bz).$$

(11.5.9)

11.6.   Discussion

Analyzing the various steps of the derivation of (11.5.8), we see that the basic result is (11.3.3). Any equation for the time history of the state of the system which leads to this type of relation will allow us to perform the same reduction of dimensionality. This observation enables us to treat certain classes of differential-difference and partial differential equations in an analogous fashion, as we shall see in Chapter 12.

11.7.   Riccati Differential Equation

Let us now use the fact that

$$\phi(z, T) = (z, S(T)z).$$

(11.7.1)

Substituting this expression in (11.5.8), we obtain the Riccati matrix differential equation

$$\frac{dS}{dT}(T) = -SK(T)K(T)'S, \qquad S(0) = B.$$

(11.7.2)

In place of the Riccati equation for a matrix of degree N, a system of $N(N + 1)/2$ quadratically nonlinear differential equations, we have an equation of degree k, a system of $k(k + 1)/2$ equations of similar nature. Despite this, there are still some questions about the feasibility of the technique which must be carefully examined; see the papers cited in the Bibliography.

EXERCISES

1.   Obtain the corresponding result for the problem posed in (11.2.4) and (11.2.5).

2. Obtain the corresponding result for the case where x and y are connected by the relation $x' = A(t)x + y,\ x(0) = c$.
3. Use the calculus of variations to obtain the corresponding Euler equation. Is it of lower degree than the Euler equation obtained for the original problem?

11.8. General Terminal Criterion

Let us turn to the case where the criterion function takes the form

$$J(x, y) = g(x(T)) + \int_0^T h(y)\, dt. \qquad (11.8.1)$$

If we assume once again that g depends only upon k components of $x(T)$, there is the possibility that we can employ the general computational approach of dynamic programming. Starting from (11.5.2), we obtain a recurrence relation

$$\psi(z, T) = \min_{y(0)} [h(y(0))\Delta + \psi(z + K(T)y(0)\Delta, T - \Delta)], \qquad (11.8.2)$$

with

$$\psi(z, 0) = g(z), \qquad (11.8.3)$$

where $\psi(z, k\Delta) \cong \phi(z, k\Delta)$. If $k \leq 3$, we can employ direct methods; if $k \geq 4$, we must employ polynomial approximation to circumvent the limitations on rapid-access storage, or alternative methods, as we have already mentioned.

11.9. Constraints

The method described in Section 11.8 is particularly valuable if there are constraints either on the allowable controls or on the allowable states of the system. Furthermore, the presence of constraints will accelerate the computation.

EXERCISE

1. Is the method applicable to discrete stochastic processes of the form $x_{n+1} = Ax_n + y_n + r_n,\ x_0 = c$, where it is desired to minimize the expected value of $J(\{x_n, y_n\}) = (x_n, Bx_n) + \sum_{k=0}^{N-1} (y_k, y_k)$?

11.10.   Successive Approximations

The standard approach of analysis to the analytic and computational treatment of equations of complicated nature is the use of successive approximations. If we cannot solve the original problem directly, we attempt to obtain its solution as the limit of a sequence of problems of a tractable type.

Abstractly, we can describe the procedure in the following way. Let the original obdurate equation be

$$T(u) = 0. \tag{11.10.1}$$

On the basis of some combination of experience and ingenuity, we write it in the form

$$L(u) = N(u), \tag{11.10.2}$$

where $L(u)$ is a tractable operator and $N(u)$ represents a stubborn term. By tractable" we mean that the inhomogeneous equation

$$L(u) = f \tag{11.10.3}$$

can readily be solved for u, $u = L^{-1}(f)$. In practice, in classical analysis, this means that L is a linear operator.

We then guess an initial approximation, u_0, and proceed to determine the sequence $\{u_n\}$, $n \geq 1$, by means of the recurrence relation

$$L(u_n) = N(u_{n-1}). \tag{11.10.4}$$

The general idea is straightforward, as all general ideas are. In any individual situation, numerous problems confront us: convergence, rate of convergence, numerical stability, choice of initial approximation, storage capacity, and so forth.

With the advent of the powerful computer (digital, analog, or hybrid) the description of "tractability" has changed drastically. In particular, the domain of tractable equations has expanded far beyond the set of linear equations of special types. Many nonlinear equations can be solved numerically, simply, quickly, and accurately.

What has become crucial then is not so much linearity as dimensionality. This means that one of the principal objectives of the algorithm of successive approximations should be to reduce the dimension of each stage of the calculation. We want to solve high-dimensional equations in terms of sequences of equations of low dimension. As usual, we are trading time for rapid-access storage capacity.

This is a part of the developing study of the decomposition of processes, a topic touched upon in Chapter Ten.

11.11.   Quadratic Case

The problem of minimizing the quadratic functional

$$J(x, y) = (x(T), Bx(T)) + \int_0^T (y, y)\, dt, \qquad (11.11.1)$$

where

$$x' = Ax + y, \qquad x(0) = c, \qquad (11.11.2)$$

leads via a direct application of dynamic programming to a Riccati equation involving a matrix of dimension N, where N is the dimension of x.

Let us explore the possibility of using successive approximations to treat the foregoing problem in terms of the limit of a sequence of problems where B has the particular form of (11.2.3). To this end, we can proceed in the following fashion. Guess initial values

$$[x_{k+1}^{(0)}(T), x_{k+2}^{(0)}(T), \ldots, x_N^{(0)}(T)], \qquad (11.11.3)$$

and replace the corresponding components of $x(T)$, $[x_{k+1}(T), \ldots, x_N(T)]$, by these values. This yields a functional $J_0(x, y)$ dependent only on $x_1(T), \ldots,$ $x_k(T)$, whose minimum can be determined following the procedure given in Section 11.7; which is to say, it can be determined by a Riccati equation for a matrix of dimension k.

Let

$$[x_1^{(0)}(T), \ldots, x_k^{(0)}(T)] \qquad (11.11.4)$$

be the values of $x_1(T), \ldots, x_k(T)$ obtained in this fashion. We now repeat this calculation, using the $N - k$ value

$$[x_1^{(0)}(T), x_2^{(0)}(T), \ldots, x_{N-k}^{(0)}(T)], \qquad (11.11.5)$$

for the first $N - k + 1$ components of $x(T)$ with $x_{N-k+1}(T), \ldots, x_N(T)$ to be determined. Carrying out the minimization, this yields new values

$$[x_{N-k+1}^{(1)}(T), \ldots, x_N^{(1)}(T)]. \qquad (11.11.6)$$

These values are used in conjunction with the previous values to provide another set of values of $N - k$ components of $x(T)$ and so on. There are many interesting aspects to this procedure, some of which we will indicate in the Exercises. Our aim is not so much to explore a particular algorithm as to indicate the kinds of novel applications of the venerable method of successive approximations which are now possible, practical, and desirable.

1. Consider the two-dimensional system

$$x_1' = a_{11}x_1 + a_{12}x_2 + y_1, \qquad x_1(0) = c_1,$$
$$x_2' = a_{21}x_1 + a_{22}x_2 + y_2, \qquad x_2(0) = c_2,$$

with the criterion function

$$J(x, y) = x_1{}^2(T) + x_2{}^2(T) + \int_0^T (y_1{}^2 + y_2{}^2)\, dt.$$

Does the foregoing procedure with $k = 1$ converge? Does it necessarily converge to the minimizing functions?

2. Does the procedure converge if we use the criterion function

$$x_1(T)^2 + bx_1(T)x_2(T) + x_2(T)^2 + \int_0^T (y_1{}^2 + y_2{}^2)\, dt_1,$$

$b \neq 0$, $|b| < 2$?

11.12.   A General Nonlinear Case

In a number of cases, as we already know, nonlinear control processes, which is to say processes where the describing equations may be nonlinear and the criterion function may be nonquadratic, can be treated as a limiting form of a succession of linear processes with quadratic criteria. Thus, for example, under appropriate assumptions concerning g, h, and k, the problem of minimizing

$$J(x, y) = k(x(T)) + \int_0^T h(x, y)\, dt \tag{11.12.1}$$

subject to

$$x' = h(x, y), \qquad x(0) = c, \tag{11.12.2}$$

can be treated.

Nonetheless, the dimension of the resultant linear system if the calculus of variations is employed, or that of the associated Riccati equation if dynamic programming is used, may still be too large for effective treatment. We have repeatedly discussed this point. In a situation of this type we may wish to employ the alternate type of successive approximations described in the previous sections. Observe, for example, that we can always convert (11.12.1) into the appropriate preliminary form by writing the integral in the form

$$\int_0^T k(x_n, y) \, dt, \qquad (11.12.3)$$

where x_n is the n-th approximation to x. A great deal remains to be done before we can use methods of this type in a routine way.

EXERCISES

1. Show that the problem of minimizing $J(x, y) = \int_0^T (y, y) + (x, Dx)] \, dt$, where $x' = Ax + y$, $x(0) = c$, and $D > 0$, can be converted into the problem of minimizing $K(x, z) = \int_0^T (z, z) \, dt + (Bx(T), x(T))$ where $x' = (A + B)x + z$, $x(0) = c$, provided that B satisfies the quadratic matrix equation $(B + A')(B + A) = D + A'A$.

2. If A is symmetric, discuss the steps involved in determining B numerically.

3. Consider the Riccati differential equation $R' = (D + A'A) - (R + A')$ $\times (R + A)$, $R(0) = 0$, where $D > 0$. What is the limiting behavior of $R(T)$ at $T \to \infty$?

4. Consider the problem of minimizing $\int_0^T (y, y) \, dt$ where $x' = Ax + y$, $x(0) = c$, $x(T) = d$, and the associated problem of minimizing $J(x, y, \lambda) = \int_0^T (y, y) \, dt + \lambda(x(T) - d, x(T) - d)$, $\lambda > 0$, with $x' = Ax + y$, $x(0) = c$. Let $x(t, \lambda)$, $y(t), \lambda)$ be the minimizing functions of the associated problem. Do $x(t, \lambda)$ and $y(t, \lambda)$ possess limits as $\lambda \to \infty$, and if so, what are they?

5. As λ increases, does the quantity $(x(t, \lambda) - d, x(T, \lambda) - d)$ decrease? What about $\int_0^T (y(t, \lambda), y(t, \lambda)) \, dt$?

6. Suppose that in place of the full condition $x(T) = d$, we had a condition involving only the first k components, $x_1(T) = d_1, \ldots, x_k(T) = d_k$. Discuss alternate ways of solving this problem computationally.

BIBLIOGRAPHY AND COMMENTS

For the material in this chapter, see

R. Bellman and R. Kalaba, "Reduction of Dimensionality, Dynamic Programming and Control Processes," *J. Basic Eng.*, March 1961, pp. 82–84.

D. C. Collins, "Reduction of Dimensionality in Dynamic Programming via the Method of Diagonal Decomposition," *JMAA*, 1969.

D. C. Collins and A. Lew, "Dimensional Approximation in Dynamic Programming by Structural Decomposition," *JMAA*, 1970, to appear.

A. Lew, "Approximation Techniques in Discrete Dynamic Programming," TR 70-10, Univ. Southern California (Dept. of Elec. Eng.), January 1970.

For some different methods, see

R. Bellman, "On the Reduction of Dimensionality for Classes of Dynamic Programming Processes," *J. Math. Anal. Appl.*, 3, 1961, pp. 358–360.

R. Bellman, "Dynamic Programming, Generalized States and Switching Systems," *J. Math. Anal. Appl.*, **12**, 1965, pp. 360–363.

11.5. For a discussion of how the explicit analytic form of the solution can be used to validate the formalism in this section, see Volume I.

See also

M. Aoki, "Control of Large Dynamic Systems by Aggregation," *IEEE Trans. Autom. Control*, June, 1968.

E. S. Lee, "Dynamic Programming, Quasilinearization and Dimensionality Difficulty," *J. Math. Anal. Appl.*, **27**, 1968, pp. 303–322.

P. Sannuto and P. V. Kokotovic, "Near Optimum Design of Linear Systems by a Singular Perturbation Method," *IEEE Trans. Autom. Control*, **AC-14**, 1969, pp. 15–21.

12

DISTRIBUTED CONTROL PROCESSES

AND THE CALCULUS OF VARIATIONS

12.1. Introduction

In this chapter we wish to examine the application of the calculus of variations to the study of control porcesses associated with systems described by state vectors of infinite dimension. Our attention will be focused on two particular processes—one described by a linear partial differential equation of parabolic type and the other described by a linear differential-difference equation.

No previous knowledge of either the theory of linear partial differential equations or linear differential-difference equations will be assumed or required. We will proceed in a reasonably free-wheeling fashion, deriving various results as we need them and referring to various sources for others. Our principal concern is to indicate the kinds of problems that can arise, some of the methods that can be employed, and the many challenging, attendant analytic and computational questions.

As we shall see, there is no difference in principle between the treatment of the finite-dimensional process of the preceding pages and the techniques we present in what follows. As usual, however, the problem of dimensionality is a serious obstacle to progress along existing lines, if we wish to obtain analytic approximations or numerical solutions. The fact that we cannot routinely apply the existing techniques applicable to low-dimensional systems is a strong incentive for continuing research.

12.2. A Heat Control Process

Let us formulate a control process involving a system described as an infinite-dimensional state vector which is analogous to the one treated in Volume I. Consider the problem of minimizing the quadratic functional

$$J(u, v) = \int_0^T \int_0^1 (q(x)u^2 + v^2) \, dx \, dt, \tag{12.2.1}$$

where u and v are connected by the linear partial differential equation

$$u_t = u_{xx} + v,$$
$$u(x, 0) = g(x), \qquad 0 < x < 1,$$
$$u(0, t) = u(1, t) = 0, \qquad t > 0. \tag{12.2.2}$$

We suppose that $g(x) \geq 0$, ensuring that J is positive definite for all $T > 0$. The function v is considered admissible if $v \in L^2$, i.e., $\int_0^T \int_0^1 v^2 \, dx \, dt < \infty$.

We will derive the Euler equation in standard fashion and then discuss some of the analytic and computational questions which arise from this procedure.

12.3. The Euler Formalism

To obtain the variational equation for v, we proceed exactly as in the finite-dimensional case. Let $\bar{u}, \bar{v}$ denote a set of functions which are candidates for the solution of the minimization problem and write

$$u = \bar{u} + \varepsilon z,$$
$$v = \bar{v} + \varepsilon w, \tag{12.3.1}$$

where ε is a scalar variable, and w and z are functions at our disposal. Then we have

$$J(u, v) = J(\bar{u}, \bar{v}) + 2\varepsilon \int_0^T \int_0^1 [q(x)z\bar{u} + w\bar{v}] \, dx \, dt + \varepsilon^2 J(z, w). \tag{12.3.2}$$

Using (12.2.2), we see that the functions z and w are related by the linear equation

$$z_t = z_{xx} + w, \qquad z(x, 0) = 0, \qquad z(0, t) = z(1, t) = 0. \tag{12.3.3}$$

The variational condition is

$$\int_0^T \int_0^1 (q(x)z\bar{u} + w\bar{v})\, dx\, dt = 0 \tag{12.3.4}$$

for "all" z and w. The validation of the end results will be discussed in the following section. Meanwhile, let us proceed formally.

Using the relation in (12.3.3) connecting z and w, this condition takes the form

$$\int_0^T \int_0^1 (q(x)z\bar{u} + (z_t - z_{xx})\bar{v})\, dx\, dt = 0. \tag{12.3.5}$$

Let us proceed to get rid of all partial derivatives by integration by parts. We have

$$\int_0^1 z_{xx}\bar{v}\, dx = z_x\bar{v}\Big]_0^1 - \int_0^1 z_x\bar{v}_x\, dx$$

$$= z_x\bar{v}\Big]_0^1 - \Big[z\bar{v}_x\Big]_0^1 + \int_0^1 z\bar{v}_{xx}\, dx. \tag{12.3.6}$$

Since $z(0, t) = z(1, t) = 0$, the second integrated term vanishes. We also have

$$\int_0^T z_t\bar{v}\, dt = z\bar{v}\Big]_0^T - \int_0^T z\bar{v}_t\, dt. \tag{12.3.7}$$

Since $z(x, 0) = 0$, we see upon collecting terms that (12.3.5) becomes

$$\int_0^T \int_0^1 z[q(x)\bar{u} + \bar{v}_t - \bar{v}_{xx}]\, dt\, dx - \int_0^T \Big[z_x\bar{v}\Big]_0^1 dt + \int_0^1 z(x, T)\bar{v}(x, T)\, dx = 0. \tag{12.3.8}$$

Since z is "arbitrary," we expect the Euler equation to be

$$v_t + v_{xx} = g(x)u, \tag{12.3.9}$$

with the terminal condition

$$v(x, T) = 0, \tag{12.3.10}$$

and the boundary conditions

$$v(0, t) = v(1, t) = 0, \qquad t > 0. \tag{12.3.11}$$

Observe that the determination of u and v using (12.2.2) and (12.3.9), (12.3.10), and (12.3.11) involves a two-point boundary-value problem, a feature which, in customary fashion, considerably complicates the task of obtaining either an analytic or a numerical solution.

EXERCISES

1. Obtain the corresponding variational equation in the more general case where $u_t = u_{xx} + v$, $u(x, 0) = g(x)$, $0 < x < 1$, $u(0, t) = u(1, t) = 0$, $t > 0$, and $J(u, x) = \int_0^T \int_0^1 h(u, u_x, v) \, dx \, dt$.

2. Obtain the corresponding equation where $u_t = u_{xx}$, $u(x, 0) = g(x)$, $u(0, t = 0$, $u(1, t) = v(t)$, and we wish to minimize $J(u, v) = \int_0^1 \int_0^0 q(x)^2 \, dx \, dt + \int_0^T v^2 \, dt$ with respect to v. This is an example of *boundary control*.

3. Obtain the variational equations for the case where $u_t = u_{xx} + v$, $u(x, 0) = g(x)$, and we wish to minimize $\lambda \int_0^T [u(0, t)^2 + u(1, t)^2] \, dt + \int_0^T \int_0^1 v^2 \, dx \, dt$.

4. Obtain the variational equations for the case where $u_t = u_{xx} + v$, $u(x, 0) = g(x)$, $u(0, t) = u(1, t) = 0$, $t > 0$, and we wish to minimize $\sum_{i=1}^K w_i u(x_i, T)^2 + \int_0^T \int_0^1 v^2 \, dx \, dt$.

12.4.   Rigorous Aspects

Which of the approaches used for the finite-dimensional case shall we pursue here? Shall we first attempt to show that the Euler equations obtained in Section 12.3 possess a unique solution and then show that this solution is the desired minimizing pair u, v, or shall we use functional analysis to derive the Euler equation?

The difficulty with the first approach lies in the fact that the Euler equations are subject to the usual two-point boundary-value conditions of the calculus of variations—one at $t = 0$ and one at $t = T$. This poses serious analytic difficulties in this infinite-dimensional space which can best be resolved by functional analysis—a method, however, which afford us little help as far as the computational aspects are concerned. As we shall now show, we can replace the two-point boundary-value problem by an integral equation of Fredholm type. This may be treated by numerical techniques, but not readily. Following this, we shall briefly discuss the use of difference approximations, which has the advantage of transforming the new problem into a type already considered.

To begin with, we want to represent the solution of

$$u_t = u_{xx} + v,$$

$$u(x, 0) = g(x), \qquad 0 < x < 1,$$

$$u(0, t) = u(1, t) = 0, \qquad t > 0, \tag{12.4.1}$$

in the form

$$u = T_1(v) + T_2(g), \tag{12.4.2}$$

where T_1 and T_2 are specific linear integral operators with the property that $v \in L^2 [(0, 1) \times (0, T)]$, $g \in L^2(0, 1)$, implies that T_1 and T_2 are L^2 over $[(0, 1)x(0, T)]$. We will now indicate how to find these operators using the Laplace transform. As always, the advantage of using integral operators resides in their closure and smoothness properties.

The original problem of minimizing

$$J(u, v) = \int_0^T \int_0^1 (q(x)u^2 + v^2) \, dx \, dt \qquad (12.4.3)$$

can thus be converted into the problem of minimizing a quadratic functional of the form

$$\int_0^T \int_0^1 q(x)[T_1(v) + h]^2 \, dx \, dt + \int_0^T \int_0^1 v^2 \, dx \, dt, \qquad (12.4.4)$$

with respect to v, where $h = T_2(g)$ is a specified function.

The functional analysis argument given in Chapter 9 of Volume I may readily be modified to yield the existence and uniqueness of a minimizing function $v \in L^2[(0, 1)x(0, T)]$ and an equation for its determination. In this case, the equation is a linear integral equation of Fredholm type, equivalent, of course, to a combination of the variational equation for v just found and the original equation connecting u and v.

We shall subsequently examine that equation in connection with computational feasibility. At the moment, let us see how to determine the operators T_1 and T_2.

12.5. Laplace Transform

There are a number of techniques available for determining the operators T_1 and T_2. One hinges upon the use of the adjoint equation; a second requires the determination of characteristic functions and characteristic values.

For the foregoing case of constant coefficients, a Laplace transform method is perhaps more convenient. Furthermore, it possesses the advantage that we can use the same method for the differential-difference equation to be considered. We shall proceed formally in what follows, referring the reader to other sources for validation of various steps. Although this validation is not difficult, it does consume a fair amount of space.

With u defined by (12.4.1), write

$$L(u) = \int_0^\infty e^{-st}u \, dt = w(x, s) = w, \qquad (12.5.1)$$

with Re $(s) > 0$. Then taking the Laplace transform of both sides of (12.4.1), we have

$$w_{xx} - sw = -L(g) - L(v), \qquad w(0, s) = w(1, s) = 0. \qquad (12.5.2)$$

Let $k(x, x', s)$ be the Green's function associated with the ordinary differential equation in (12.5.2). Then (12.5.2) yields

$$w = \int_0^1 k(x, x', s)[L(g) + L(v)]\,dx \,. \qquad (12.5.3)$$

Since g depends only on x, we have

$$L(g) = g/s. \qquad (12.5.4)$$

Hence, (12.5.3) yields

$$w = \int_0^1 \frac{k(x, x', s)}{s} g(x')\,dx' + \int_0^1 k(x, x', s)L(v)\,dx'. \qquad (12.5.5)$$

To obtain an expression for $u = L^{-1}(w)$, we invoke the convolution theorem of Emil Borel, namely

$$L\left(\int_0^t u_1(t - t_1)u_2(t_1)\,dt_1\right) = L(u_1)L(u_2). \qquad (12.5.6)$$

Setting

$$K(x, x', t) = L^{-1}(k(x, x', s)), \qquad (12.5.7)$$

we obtain from (12.5.5) the expression

$$u = \int_0^1 \left(\int_0^t K(x, x', t_1)\,dt_1\right)g(x')\,dx' + \int_0^1 \left[\int_0^t K(x, x', t - t_1)v(x', t_1)\,dt_1\right] dx'. \qquad (12.5.8)$$

Let us note for future reference in connection with a terminal control process that this yields

$$u(x, T) = \int_0^1 \left(\int_0^T K(x, x', t_1)\,dt_1 g(x')\,dx'\right.$$
$$+ \int_0^1 \left[\int_0^T K(x, x', T - t_1)v(x', t_1)\,dt_1\right] dx'$$
$$= T_2(g) + T_1(v). \qquad (12.5.9)$$

Once the relations have been derived in this fashion, they can be verified directly.

EXERCISES

1. Use the method of separation of variables to show that $K(x, x', t) = \sum_{n=1}^{\infty} \exp{-(n^2\pi^2 t)} \sin n\pi x \sin n\pi x'$, $t > 0$, $0 \le x, x' \le 1$.
2. Does the function $K(x, x', t)$ belong to $L^2[(0, 1) \times (0, T)]$?
3. Does $\int_0^t K(x, x', t_1) \, dt_1$ belong to L^2 in this region?
4. Does $\int_0^1 [\int_0^t K(x, x', t - t_1) v(x', t_1) \, dt_1] \, dx_1$ belong to L^2 in this region if v does?
5. Calculate the Green's function $k(x, x', s)$ and use the complex inversion formula $L^{-1}(k) = 1/2\pi i \int_C e^{st} k(x, x', s) \, ds$, with C a suitable contour to obtain K, as given in Exercise 1.

12.6. An Integral Equation

Let us return briefly to the system

(a) $u_t = u_{xx} + v$,

   $u(x, 0) = g(x)$, $0 < x < 1$,

   $u(0, t) = u(1, t) = 0$, $t > 0$,

(b) $v_t + v_{xx} = q(x)u$,

   $v(x, T) = 0$,

   $v(0, t) = v(1, t) = 0$, $t > 0$.   (12.6.1)

Consider instead the initial-value system where the equations in (12.6.1b) are replaced by

$$v_t + v_{xx} = q(x)u,$$
$$v(x, 0) = w(x),   0 < x < 1,   \qquad (12.6.2)$$
$$v(0, t) = v(1, t) = 0,   t > 0,$$

where $w(x)$ is a function to be determined.

   Using Laplace transforms as above, this new system, (12.6.1a) and (12.6.2), may be solved to yield

$$u = T_{11}(g) + T_{12}(w),$$
$$v = T_{21}(g) + T_{22}(w), \qquad (12.6.3)$$

where T_{ij} are linear integral operators similar in form to T_2. Setting $v(x, T) = 0$, we obtain the desired Fredholm equation for w.

12.7.   The Variational Equation

Let us return to a rigorous derivation of the variational equation. We drop the subscript and write $T(c)$ in place of $T_1(v)$ in (12.5.9). Our objective then is to minimize the quadratic functional

$$J(v) = \int_0^T \int_0^1 q(x)(T(v) + h)^2 \, dx \, dt + \int_0^T \int_0^1 v^2 \, dx \, dt \qquad (12.7.1)$$

over the class of functions v belonging to L^2 over the region R defined by $0 \le x \le 1$, $0 \le t \le T$. The functional analysis argument used in Chapter 9 of Volume I shows that a unique minimizing v exists and is specified by the unique solution of the variational equation

$$T^*(q[T(v) + h]) + v = 0. \qquad (12.7.2)$$

Here T^* is the transformation adjoint to T over R. This equation is a linear integral equation of Fredholm type which possesses a unique solution. Although we now have a firm analytic basis, considerable thought must be devoted to the question of computational feasibility.

EXERCISES

1.   Show that (12.7.2) is equivalent to the pair of equations for u and v obtained in Section 12.3.
2.   Show that the Fredholm integral equation has a symmetric kernel.

12.8.   Computational Approaches

We can, if we wish, return to the pair of linear partial differential equations given in Section 12.14. Discretization of the usual type will yield a system of linear differential equations subject to two-point boundary conditions, which is a type of problem we have already discussed. We shall now discuss two types of discretization procedures.

We can also attack the integral equation of (12.7.2) directly by using quadrature techniques, thereby obtaining an approximating system of linear algebraic equations. The matrix of the system will be symmetric, which is a distinct computational advantage.

12.9.   Modified Liouville–Neumann Solution

In some cases, it may be feasible to approach (12.7.2) directly using the Liouville-Neumann iterative solution. Let us write the equation in the form

$$v + Av = w, \qquad (12.9.1)$$

defining in this way the operator A and the function w, and introduce the nonnegative parameter λ,

$$v + \lambda A v = w. \tag{12.9.2}$$

The Liouville-Neumann solution is then

$$v = w - \lambda A w + \lambda^2 A^2 w - \cdots. \tag{12.9.3}$$

It is clear from the foregoing discussion that (12.9.2) possesses a unique solution for any $\lambda \geq 0$, since introduction of λ changes the operator T in an inessential fashion. Despite this, (12.9.3) will converge only for a restricted range of values of λ, $0 \leq \lambda < \lambda_0$. In particular, (12.9.3) may not converge for $\lambda = 1$, the value of particular interest, despite the fact that (12.9.1) possesses an unique solution. If it does converge, it may converge rather slowly.

It turns out, in view of the nature of A as a positive definite operator, that we can always introduce a change of parameter,

$$\mu = \frac{\lambda}{\lambda + b}, \tag{12.9.4}$$

where $b > 0$ is determined by the spectral radius of A, with the property that v written as a power series in μ converges for *all* $\lambda \geq 0$. The details, together with some examples of the use of this type of transformation, will be found in Chapter 9 of Volume I.

The determination of b is not an easy matter. However, in this case it is relatively easy to obtain good approximations to its value since A is a positive operator in addition to being positive definite. Any reasonable estimate of b enables us to extend the usefulness of the series in (12.9.3).

EXERCISE

1. Obtain the power series in μ.

12.10. Use of Difference Approximations

If we shall ultimately use a digital computer to obtain numerical results, why not develop a formulation in discrete terms from the very beginning? One answer is that the preparatory steps may be more easily carried through on the grounds of both esthetics and simplicity. Let us, however, sketch one approach in discrete terms.

We begin by replacing the second derivative by the second difference,

$$u_{xx} \cong \frac{u(x + \Delta) + u(x - \Delta) - 2u(x)}{\Delta^2}. \tag{12.10.1}$$

The variable x is allowed only the discrete values Δ, 2Δ, $\ldots$, $(N-1)\Delta$, with $N\Delta = 1$. The choice of Δ is critical—too large and we founder on the shoals of inaccuracy, too small and we are swamped by dimensionality. Here the concept of "deferred passage to the limit" can play a vital role in allowing us to obtain high precision without an excessive dimensionality or time cost.

Write

$$u(k\Delta, t) = u_k(t), \qquad k = 0, 1, 2, \ldots, N,$$

$$v(k\Delta, t) = v_k(t), \qquad k = 1, 2, \ldots, N-1, \qquad (12.10.2)$$

$$g(k\Delta) = g_k, \qquad k = 0, 1, \ldots, N$$

The equation of (12.4.1) is replaced by the finite system

$$u_k' = (u_{k+1} + u_{k-1} - 2u_k)/\Delta^2 + v_k, \qquad k = 1, 2, \ldots, N-1,$$

$$u_k(0) = g_k. \qquad (12.10.3)$$

Let us take $g_0 = g_n = 0$ to preserve continuity. Finally, the original criterion function may be replaced by the finite sum

$$J(\{u_k, v_k\}) = \sum_{k=0}^{N-1} \Delta \int_0^T (q_k u_k^2 + v_k^2)\, dt. \qquad (12.10.4)$$

The resultant finite-dimensional control process may be treated either by the calculus of variations or by dynamic programming.

EXERCISE

1. What are advantages and disadvantages involved in using a difference approximation of the form

$$u(x, t + \Delta^2/2) = \frac{1}{2}[u(x + \Delta, t) + u(x - \Delta, t)] + v(x, t)\Delta^2/2,$$

where $x = 0, \Delta, 2\Delta, \ldots, N\Delta = 1$ as before and $t = 0, \Delta^2/2, \Delta^2, \ldots$, and eliminating all derivatives?

12.11. Generalized Quadrature

Let us recall that by quadrature we mean an approximate evaluation of an integral of the form

$$\int_a^b f(t)\, dt \cong \sum_{i=1}^N w_i f(t_i). \qquad (12.11.1)$$

A particularly important example of this is the quadrature formula of Gauss,

$$\int_{-1}^{1} f(t)\, dt \cong \sum_{i=1}^{N} w_i\, f(t_i),\tag{12.11.2}$$

where the weights w_i and the quadrature points t_i are chosen so that the relation is exact for polynomials of degree $2N - 1$ or less. The advantage of this type of approximation over approximations at regularly spaced points is that fewer evaluations of f are required for the same degree of accuracy. This is an essential point in our constant battle against dimensionality.

An interesting extension of the classical concept of quadrature is the representation of any linear functional in the form of (12.11.2). In particular, in connection with the heat control process just considered, we want to write

$$u''(x_i) \cong \sum_{j=1}^{T} a_{ij}\, u(x_j)\tag{12.11.3}$$

for a particular set of points $x_1, x_2, \ldots, x_N$. The coefficients a_{ij} may in this case be determined by the condition that (12.11.3) is exact for polynomials of degree N or less which are constrained by the conditions $u(0) = u(1) = 0$. As in the case of conventional quadrature, there are a number of alternate determinations.

With these preliminaries, let us return to the partial differential equation of (12.6.1). Using (12.11.3), we obtain the approximating system

$$u_i' = \sum_{j=1}^{N} a_{ij} u_j + v_i,\tag{12.11.4}$$

$$u_i(0) = g_i, \qquad i = 1, 2, \ldots, N,$$

where

$$u_i = u(x_i, t), \qquad v_i = v(x_i, t), \qquad g_i = g(x_i).\tag{12.11.5}$$

Using (12.11.2), suitably modified to hold for $[0, 1]$ in place of $[-1, 1]$, we obtain the associated criterion function

$$J(\{u_i, v_i\}) = \sum_{i=1}^{N} w_i\, q(x_i) \int_{0}^{T} u_i^2\, dt + \sum_{i=1}^{N} w_i \int_{0}^{T} v_i^2\, dt.\tag{12.11.6}$$

The problem of minimizing J subject to (12.11.4) is one we have treated in detail in Volume I.

12.12.   Orthogonal Expansion and Truncation

Let $\{\phi_n(x)\}$ be an orthonormal sequence over $[0, 1]$ and write

$$u(x, t) \sim \sum_{n=1}^{\infty} u_n(t)\phi_n(x),$$

$$v(x, t) \sim \sum_{n=1}^{\infty} v_n(t)\phi_n(x). \qquad (12.12.1)$$

By means of a particular choice of the $\phi_n(x)$, we can replace the original partial differential equation by an infinite system of ordinary differential equations, and we can replace the criterion functional by a quadratic functional in the u_n and v_n. Truncating, i.e., setting

$$u_n = 0, \qquad n > N, \qquad (12.12.2)$$

we obtain in this way an approximating finite-dimensional control process.

To illustrate this approach, one of the fundamental approaches of mathematical physics, let us set

$$\phi_n(x) = \sqrt{2} \sin n\pi x. \qquad (12.12.3)$$

Then the partial differential equation is " diagonalized,"

$$u_n'(t) = -n^2\pi^2 u_n(t) + v_n(t), \qquad n = 1, 2, \ldots,$$

$$u_n(0) = \sqrt{2} \int_0^1 g(x) \sin n\pi x \, dx, \qquad (12.12.4)$$

and the criterion function takes the form

$$\int_0^T \int_0^1 [g(x)u^2 + v^2] \, dx \, dt$$

$$= \int_0^T \int_0^1 \left[g(x)\left(\sum_{n=1}^{\infty} u_n(t)\phi_n(x) \right)^2 + \left(\sum_{n=1}^{\infty} v_n(t)\phi_n(x) \right)^2 \right] dx \, dt$$

$$\cong \int_0^T \int_0^1 \left[g(x)\left(\sum_{n=1}^{N} u_n(t)\phi_n(x) \right) \right]^2 dx \, dt + \int_0^T \left(\sum_{n=1}^{N} v_n^2(t) \right) dt, \quad (12.12.5)$$

upon introducing truncation in the straightforward fashion indicated above. In (12.12.4), we let n assume only the values $1, 2, \ldots, N$. We thus obtain the desired-finite dimensional process which can be treated by available methods. We would expect that nonlinear extrapolation would play a significant role in deducing the limiting behavior as $N \to \infty$ from the behavior for moderate N.

12.13. Differential-difference Equation

As a second example of an infinite-dimensional control process, let us consider a system described by a differential-difference equation. Physical processes in which time lags must be taken into account generate equations of this type. Frequently, the time lag is a consequence of the control process, itself, a point discussed in Chapter 14.

Let us then examine the question of minimizing the quadratic functional

$$J(u, v) = \int_1^T (u^2 + v^2)\, dt, \tag{12.13.1}$$

where u and v are connected by the relation

$$
\begin{aligned}
u'(t) &= a_1 u(t) + a_2 u(t-1) + v(t), \quad t \geq 1, \\
u(t) &= g(t), \quad 0 \leq t \leq 1.
\end{aligned}
\tag{12.13.2}
$$

We take g continuous and allow v to range over $L^2(0, T)$.

12.14. The Euler Equation

Prior to any rigorous discussion of existence and uniqueness of a minimizing function v, let us derive the Euler equation. As usual, write

$$u = \bar{u} + \varepsilon w, \tag{12.14.1}$$

$$v = \bar{v} + \varepsilon z,$$

where $\bar{u}, \bar{v}$ are the minimizing pair. Then

$$J(u, v) = J(\bar{u}, \bar{v}) + 2\varepsilon \int_1^T (\bar{u}w + \bar{v}z)\, dt + \varepsilon^2 J(w, z), \tag{12.14.2}$$

while (12.13.2) yields

$$w'(t) = a_1 w(t) + a_2 w(t-1) + z(t). \tag{12.14.3}$$

The variational condition is

$$\int_1^T (\bar{u}w + \bar{v}z)\, dt = 0. \tag{12.14.4}$$

Using (12.14.3) to eliminate z, this may be written

$$\int_1^T (\bar{u}w + \bar{v}[w'(t) - a_1 w(t) - a_2 w(t-1)])\, dt = 0. \tag{12.14.5}$$

Let us note

$$\int_1^T \bar{v}w'(t)\,dt = \bar{v}w\Big]_1^T - \int_1^T \bar{v}'w\,dt,$$

$$\int_1^T \bar{v}w(t-1)\,dt = \int_0^{T-1} \bar{v}(t+1)w(t)\,dt. \qquad (12.14.6)$$

Since $w(1) = 0$, there is no contribution to the integrated term at $t = 1$. Using the results in (12.14.5), (12.14.6), yields

$$\int_1^{T-1} w(t)[\bar{u}(t) - \bar{v}'(t) - a_1\bar{v}(t) - a_2\,\bar{v}(t+1)]\,dt$$

$$+ \int_{T-1}^T w(t)[\bar{u}(t) - v'(t) - a_1\bar{v}(t)]\,dt + \bar{v}(T)w(T) = 0. \quad (12.14.7)$$

Hence, we obtain the equations (dropping the overbars)

$$u(t) - v'(t) - a_1 v(t) - a_2\,v(t+1) = 0, \qquad 1 \le t \le T-1,$$
$$u(t) - v'(t) - a_1 v(t) = 0, \qquad T-1 \le t \le T, \quad (12.14.8)$$
$$v(T) = 0.$$

What is interesting about the set (12.14.2) and (12.14.8) is that they constitute a pair of *two-interval* boundary-value equations. Equations of this nature have not been studied to any extent, either analytically or computationally.

12.15.   Discussion

If we could establish in some fashion the existence of a pair of functions satisfying (12.13.2) and (12.14.8), we could in a familiar fashion use (12.14.2) to demonstrate that these functions furnish the absolute minimum of $J(u, v)$; thus there is uniqueness of solution. However, it does not appear easy to establish the existence of a solution without the use of functional analysis. Consequently, we shall abandon the foregoing route and follow a path akin to that for the partial differential equation employing some functional analysis from the very beginning

12.16.   Laplace Transform

Starting with

$$u'(t) = a_1 u(t) + a_2\,u(t-1) + v(t), \quad t \ge 0,$$
$$u(t) = g(t), \quad -1 \le t \le 0 \qquad (12.16.1)$$

(where we have shifted t one unit for convenience), we have

$$L(u') = a_1 L(u) + a_2 L(u(t-1)) + L(v). \qquad (12.16.2)$$

Since

$$L(u') = -u(0) + sL(u)$$
$$= -g(0) + sL(u), \qquad (12.16.3)$$

and

$$L(u(t-1)) = \int_0^\infty u(t-1)e^{-st}\, dt$$

$$= e^{-s} \int_{-1}^\infty u(t)e^{-st}\, dt$$

$$= e^{-s} \int_{-1}^0 u(t)e^{-st}\, dt + e^{-s}L(u)$$

$$= e^{-s} \int_{-1}^0 g(t)e^{-st}\, dt + e^{-s}L(u), \qquad (12.16.4)$$

we have finally

$$(s - a_1 - a_2 e^{-s})L(u) = g(0) + e^{-s} \int_{-1}^0 g(t)e^{-st}\, dt + L(v). \quad (12.16.5)$$

Hence, setting

$$k(t) = L^{-1}\left(\frac{1}{s - a_1 - a_2 e^{-s}}\right), \qquad (12.16.6)$$

we obtain the result

$$u = g(0)k(t) + \int_{-1}^0 k(t - t_1 - 1)g(t_1)\, dt_1 + \int_0^t k(t - t_1)v(t_1)dt_1. \quad (12.16.7)$$

We write this relation in the form

$$u = h + T(v), \qquad (12.16.8)$$

where the function h and the operation $T(v)$ are defined by (12.16.7).
For future reference, note that

$$u(T) = h(T) + \int_0^T k(T - t_1)v(t_1)\, dt_1. \qquad (12.16.9)$$

EXERCISE

1. Using the complex inversion formula and residue evaluation, obtain formally an expansion for $k(t)$ of the form $\sum_i a_i \exp(\lambda_i t)$.

12.17.   Integral Equation

Using the foregoing techniques we can readily derive an integral equation equivalent to the original two-interval problem. Consider the initial-value problem

$$u'(t) = a_1 u(t) + a_2 u(t - 1) + v(t), \qquad T \geq t \geq 1,$$

$$u(t) = g(t), \qquad 0 \leq t \leq 1,$$

$$v'(t) = -a_1 v(t) - a_2 v(t + 1) + u(t), \qquad 1 \leq t \leq T - 1,$$

$$v(t) = w(t), \qquad 0 \leq t \leq 1,$$

where $w(t)$ is a function to be determined.

From this, we obtain an expression for $v(t)$, $t - 1 \leq t \leq T$, as a linear integral operator involving w. On the other hand, v satisfies the equation

$$u(t) - v'(t) - a_1 v(t) = 0, \qquad T - 1 \leq t \leq T,$$

$$v(T) = 0.$$

Equating the two expressions for v yields the desired integral equation.

EXERCISE

1.   Obtain this integral equation.

12.18.   Variational Equation

The original variational problem may be transformed into the problem of minimizing the quadratic functional

$$J(v) = \int_1^T [(h + T(v))^2 + v^2] \, dt, \tag{12.18.1}$$

where h and $T(v)$ are given by (12.16.7). The general result of Volume I, Chapter 9, may now be employed to obtain the variational equation and to demonstrate the existence and uniqueness of solution. This provides a firm basis for the more difficult part of the investigation, the derivation of a numerical solution.

EXERCISE

1.   Obtain the variational equation and demonstrate that it is equivalent to (12.13.2) and (12.14.8).

12.19. Discretization

It is easy to employ discretization to convert (12.13.2) into a finite-dimensional equation. Write

$$\frac{u(t + \Delta) - u(t)}{\Delta} \cong a_1 u(t) + a_2 u(t - N\Delta) + v(t), \qquad (12.19.1)$$

$t = 1, 1 + \Delta, 1 + 2\Delta, \ldots, (N\Delta = 1)$. This is a high-order difference equation. Similarly, we replace the integral $\int_1^T (u^2 + v^2)\, dt$ by the finite sum

$$\sum_{k=0}^{N-1} \{u(k\Delta)^2 + v(k\Delta)^2\}\Delta. \qquad (12.19.2)$$

Problems of this nature were discussed in Volume I.

EXERCISE

1. How large a value of N can we handle routinely?

12.20. Discussion

In the preceding pages we have shown by example that there are no conceptual problems in constructing an analytic theory of distributed control processes along conventional lines. The dimensionality barrier to numerical solution, however, forces the construction of new types of approximation. At the present time very little is known.

Miscellaneous Exercises

1. Show that the problem of minimizing

$$J(u) = \int_0^T \left(\sum_{j=1}^N g_j(u'(t - t_j), u(t - t_j), t - t_j) \right) dt,$$

with $u(t) = h(t)$, $-1 \leq t \leq 0$, $0 = t_0 < t_1 < t_2 < \cdots < t_N = 1$, can be transformed into a problem not involving lags at the expense of introducing some internal discontinuities. (Brown).

2. Minimize

$$J(u) = \int_0^T [u'(t)^2 + u(t - (1/2))^2 + 3u(t - 1)^2]\, dt,$$

with $u(t) = g(t)$, $-1 \leq t \leq 0$.

3.   Consider the problem of minimizing

$$J(u) = g(u(T)) + \int_1^T h(u)\, dt$$

with respect to v where $u'(t) = a_1 u(t) + a_2 u(t-1)$, $t \geq 1$, $u(t) = v(t)$, $0 \leq t \leq 1$, and, in particular, the two cases

(a) $g = 0$, $h(u) = u^2$,

(b) $g = u^2$, $h(u) = 0$.

BIBLIOGRAPHY AND COMMENTS

12.1.   Detailed discussions of the calculus of variations applied to distributed control processes may be found in

H. T. Banks, "Variational Problems Involving Functional Differential Equations," *J. Siam Contr.*, **7**, 1969, pp. 1–17.

A. G. Butkowski, *Distributed Control Processes*, Elsevier, New York, 1970.

A. G. Butkowski, A. I. Egonov, and K. A. Lurie, "Optimal Control of Distributed Systems (A Survey of Soviet Publications), " *J. SIAM Comtrol*, **6**, 1968, pp. 437–476.

D. K. Hughes, "Variational and Optimal Control Problems with Delayed Argument," *J. Optim. Theory Appl.*, **2**, 1968, pp. 1–14.

S. Hinatsuka and A. Ichikawa, "Optimal Control of Systems with Transportation Lags," *IEEE Trans. Autom. Contr.*, **AC-14**, 1969, pp. 237–247.

(This contains a discussion of variable lags as well as contrast lags.)

N. N. Krasovskii, "On the Analytical Construction of an Optimal Control in Systems with Time Lags," *J. Appl. Math. Mech.*, **26**, 1962, pp. 50–67.

R. Lattes and J. L. Lions, *The Method of Quasi-reversibility and Partial Differential Equations*, Elsevier, New York, 1969.

See also

L. E. El'sgol'c, "Variational Problems with a Delayed Argument," *Amer. Math. Soc. Translat.*, Ser. 2, **16**, 1960.

Control processes involving the heat equation, or diffusion equation, are closely associated with stochastic control theory. See, for example,

H. J. Kushner, *Stochastic Stability and Control*, Academic Press, New York, 1967.

W. M. Wonham, *Random Differential Equations in Central Theory: Probabilistic Methods in Applied Mathematics*, Vol. 2, Academic Press, 1970, pp. 131–212.

12.4.   This approach based upon the use of functional analysis was first presented in

R. Bellman, I. Glicksberg, and O. Gross, *Some Aspects of the Mathematical Theory of Control Processes*, The RAND Corp., R-313, 1958.

See also Chapter 9 of Volume I.

12.6.   This technique of converting a boundary-value problem into an integral equation is a standard technique in the theory of partial differential equations.

12.9.   See

R. Bellman, *Methods of Nonlinear Analysis*, Academic Press, New York, 1970.
  12.10. See

R. Bellman and R. Kalaba, "New Methods for the Solution of Partial Differential Equations," *Nonlinear Partial Differential Equations*, Academic Press, New York, 1967, pp. 43–54.

for further discussion.
  12.11. For an application of this generalized quadrature technique, see

R. Bellman and R. Kalaba, "On a New Approach to the Numerical Solution of a Class of Partial Differential-integral Equations of Transport Theory," *Proc. Nat. Acad. Sci. USA*, **54**, 1965, pp. 1293–1296.

For a discussion of rigorous aspects, see

H. R. Keller, "On the Pointwise Convergence of the Discrete-ordinate Method," *J. Soc. Indus. Appl. Math.*, **8**, 1960, pp. 560–567.
  12.12. There are a number of interesting questions associated with truncation procedure. See

R. Bellman, "On the Validity of Truncation for Infinite Systems of Ordinary Differential Equations Associated with Nonlinear Partial Differential Equations," *Math. & Phys. Sci.*, **1**, 1967, pp. 95–100.
  12.13. For the requisite results in the theory of differential-difference equations, see

R. Bellman and K. L. Cooke, *Differential-difference Equations*, Academic Press, New York, 1963.

See also

J. D. R. Kramer, Jr., "On Control of Linear Systems with Time Lags," *Inform. and Control*, **3**, 1960, pp. 299–326.

For a description of some of the ways in which control processes of this nature can arise in economic theory, see

Z. Grihches, "Distributed Lags: A Survey," *Econometrica*, **35**, 1967, pp. 16–49.
  12.20. See
R. Bellman, "A New Type of Approximation leading to Reduction of Dimensionality in Control Processes," *JMAA*, **27**, 1969, pp. 454–459.

13

DISTRIBUTED CONTROL PROCESSES

AND DYNAMIC PROGRAMMING

13.1.   Introduction

In this chapter we wish to discuss some applications of the theory of
dynamic programming to the study of distributed control processes and to
the study of partial differential and integral equations—equations which arise
from variational processes. We shall find, as in the previous chapter, that
there is relatively little difficulty involved in extending the basic ideas used in
the infinite-dimensional case as far as constructing an analytic framework is
concerned. The significant obstacle to progress is dimensionality, an obstacle
which forces us to develop some new methods of approximation. Some of
these methods will be briefly discussed.

13.2.   Formulation of a General Control Process

In the previous chapter we considered the application of the methodology
of the calculus of variations to the minimization of the functional

$$J(u, v) = \int_0^T \int_0^1 (u^2 + v^2) \, dx \, dt, \qquad (13.2.1)$$

where u and v are connected by the equation

$$u_t = u_{xx} + v,$$

$$u(x, 0) = g(x), \qquad\qquad 0 < x < 1, \qquad\qquad (13.2.2)$$

$$u(0, t) = u(1, t) = 0, \qquad t > 0,$$

and to an analogous problem for the case where u is ruled by a differential-difference equation.

Stimulated by particular questions of this genre, let us pose the more general problem of minimizing the functional

$$J(u, v) = \int_0^T \left(\int_R g(u, v) \, dA \right) dt, \qquad\qquad (13.2.3)$$

where $u = u(p)$, $v = v(p)$, $p \in R$, with R a region bounded by B, and u and v are related by the equation

$$u_t = L(u) + h(u, v), \qquad t > 0,$$

$$u_{t=0} = c(p), \qquad\qquad p \in R, \qquad\qquad (13.2.4)$$

$$u_{p \in B} = g(p), \qquad\qquad t > 0,$$

$L(u)$ is a linear operator with properties we shall bequeath as we need them.

13.3.   Multistage Decision Process

As in the finite-dimensional case we view this variational problem as a multistage decision process. The state vector may be taken to be the function $c(p)$ with the decision variable the function v.

We introduce the return functional

$$f(c(p), T) = \min_v J(u, v), \qquad\qquad (13.3.1)$$

and wish to obtain a functional equation for f which will yield the optimal policy.

13.4.   Functional Derivative

The functional equation for $f(c(p), T)$ will involve a conventional partial derivative with respect to T as well as a new concept, a *functional derivative* with respect to $c(p)$. Let us briefly discuss this important tool of functional analysis. Consider the difference quotient

$$D(f, w, \Delta) = \frac{f(c(p) + \Delta w(p), T) - f(c(p), T)}{\Delta}, \qquad (13.4.1)$$

and suppose that the limit of this quantity exists as $\Delta \to 0$. Let us further assume that the limit is a linear functional of w having the form

$$\int_R \frac{\delta f}{\delta c} w \, dA, \qquad (13.4.2)$$

defined in some interval $0 \le T \le T_0$, in the region R, and for some class of functions embracing c and w. The coefficient of w is called the *functional partial derivative*.

EXERCISES

1.   If $f(c(p), T) = \int_R k(p, T)c(p) \, dA$, determine $\delta f/\delta c$.
2.   Determine $\delta f/\delta c$ if $f(c(p), T) = \int_R \int_{R_1} (k(p, p_1, T)c(p)c(p_1) \, dA \, dA_1$.

13.5.   Formal Derivation of Functional Equation

Let us now parallel the procedure of Chapter 6. We suppose that L is a linear operator such that (13.2.4) can be converted into the relation

$$u(p, t) = F_1(c(p), t) + F_2(h(u, v), t), \qquad (13.5.1)$$

with F_1 and F_2 linear operators determined by L.

For small t, say $t = \Delta$, we further suppose that

$$F_1(c(p), \Delta) = c(p) + F_3(c(p))\Delta + \cdots,$$
$$F_2(h(u, v), \Delta) = h(u, v)\Delta + \cdots$$
$$= h(c(p), v)\Delta + \cdots, \qquad (13.5.2)$$

where the three dots indicate terms of order Δ^2. Writing

$$\int_0^T = \int_0^\Delta + \int_\Delta^T, \qquad (13.5.3)$$

we have, proceeding completely formally,

$$f(c(p), T) = \min_v \left[\Delta \int_R g(c(p), v) \, dA \right.$$
$$\left. + f(c(p) + F_3(c(p))\Delta + h(c,(p) v)\Delta, T - \Delta) \right] + o(\Delta). \quad (13.5.4)$$

Using the functional partial derivative described in Section 13.4, we may write

$$f(c(p) + F_3(c(p))\Delta + h(c(p), v)\Delta, T - \Delta)$$

$$= \Delta \int_R \left[F_3(c(p)) + h(c(p), v), \frac{\delta f}{\delta c} \right] dA - \Delta \frac{\partial f}{\partial T} + o(\Delta). \quad (13.5.5)$$

Hence, (13.5.4) becomes in the limit as $\Delta \to 0$

$$\frac{\partial f}{\partial T} = \min_v \left[\int_R g(c(p), v) \, dA + \int_R \left[F_3(c(p)) + h(c(p), v), \frac{\delta f}{\delta c} \right] dA \right]. \quad (13.5.6)$$

The initial condition is

$$f(c(p), 0) = 0. \qquad (13.5.7)$$

13.6. Discussion

The minimization with respect to v in some cases may be carried out by calculus. In general, it will require the calculus of variations. When the minimization operation is performed, we obtain an equation of the form

$$\frac{\partial f}{\partial T} = \phi\left(\frac{\delta f}{\delta c}\right), \qquad f(c(p), 0) = 0, \qquad (13.6.1)$$

a partial differential-functional equation.

13.7. Quadratic Criteria

The procedure is readily carried through in detail for the case where the describing equation is

$$u_t = L(u) + v, \qquad (13.7.1)$$

and the criterion functional is quadratic in u and v. References to a good deal of work of this nature will be found at the end of this chapter.

13.8. Integral Equations

In order to indicate the type of result that can be expected, let us consider the linear integral equation

$$u(t) + v(t) + \int_a^T k(t, s)u(s) \, ds = 0, \qquad 0 \le a \le T, \qquad (13.8.1)$$

the "Euler equation" associated with the minimization of the quadratic functional

$$J(u) = \int_a^T u^2(t)\, dt + 2\int_a^T u(t)v(t)\, dt + \int_a^T \int_a^T k(t, s)u(t)u(s)\, dt\, ds, \quad (13.8.2)$$

over $u \in L^2(a, T)$.

We suppose that $k(t, s)$ is a symmetric kernel with the property that the quadratic term

$$\int_a^T \int_a^T k(t, s)u(t)u(s)\, dt\, ds + \int_a^T u^2(t)\, dt \qquad (13.8.3)$$

is positive definite, i.e., bounded from below by $b \int_a^T u^2(t)\, dt$ for some $b > 0$ and $0 \le a \le T$. Then it is easy to demonstrate that the convex functional $J(u)$ possesses an absolute minimum furnished by the unique solution of (13.8.1). This solution may be represented in the form

$$u(t) = -v(t) + \int_a^T q(t, s, a)v(s)\, ds, \qquad (13.8.4)$$

where the kernel $q(t, s, a)$ is called the *Fredholm resolvent*. Our objective is to obtain a nonlinear equation of Riccati type for $q(t, s, a)$.

13.9. Expression for min $J(u)$

We may write

$$J(u) = \int_a^T \left(u(t) + v(t) + \int_a^T k(t, s)u(s)\, ds\right) u(t)\, dt + \int_a^T u(t)v(t)\, dt$$

$$= \int_a^T u(t)v(t)\, dt \qquad (13.9.1)$$

for the minimizing u. Using the expression for u in (13.8.4), we may write

$$\min_u J(u) = \int_a^T \left(-v(t) - \int_a^T q(t, s, a)v(s)\, ds\right) v(t)\, dt$$

$$= -\int_a^T v^2(t)\, dt + \int_a^T \int_a^T q(t, s, a)v(s)v(t)\, ds\, dt$$

$$= f(v(t), a). \qquad (13.9.2)$$

13.10. Functional Equation for $f(v, a)$

Let us write

$$
J(u) = \int_a^{a+s} u^2(t) \, dt + 2 \int_a^{a+s} u(t)v(t) \, dt + 2 \int_a^{a+s} \int_{a+s}^T k(t, s)u(t)u(s) \, dt \, ds
$$
$$
+ \int_{a+s}^T u^2(t) \, dt + 2 \int_{a+s}^T u(t)v(t) \, dt + \int_{a+s}^T \int_{a+s}^T k(t, s)u(t)u(s) \, dt \, ds
$$
$$
+ \int_a^{a+s} \int_a^{a+s} k(t, s)u(t)u(s) \, dt \, ds. \tag{13.10.1}
$$

Regarding s as an infinitesimal, it follows in the usual fashion that

$$
f(v(t), a) = \min_u J(u) = \min_{u(a)} [s[u^2(a) + 2u(a)v(a)]
$$
$$
+ f(v(t) + su(a)k(a, t), a + s)] + o(s). \tag{13.10.2}
$$

Write, as above,

$$
L(w(t), a) = \lim_{s \to 0} \left[\frac{f(v(t) + sw(t), a) - f(v(t), a)}{s} \right]. \tag{13.10.3}
$$

Then the limiting form of (13.10.2) is

$$
0 = \min_{u(a)} \left[u^2(a) + 2u(a)v(a) + u(a)L(k(a, t), a) + \frac{\partial}{\partial a} \right]. \tag{13.10.4}
$$

The minimum is attained at

$$
u(a) = -v(a) - \frac{L(k(a, t), a)}{2}, \tag{13.10.5}
$$

yielding the quadratically nonlinear equation

$$
\frac{\partial f}{\partial a} = (v(a) + (L(k(a, t), a)/2))^2. \tag{13.10.6}
$$

13.11. The Form of $L(k(a, t), a)$

From the expression for $f(v, a)$, we readily see that

$$
L(w, a) = -2 \int_a^T v(t)w(t) \, dt + 2 \int_a^T \int_a^T q(t, s, a)v(t)w(s) \, dt \, ds. \tag{13.11.1}
$$

Thus,

$$L(k(a, t), a) = -2 \int_a^T v(t)k(a, t)\, dt + 2 \int_a^T \int_a^T q(t, s, a)k(a, s)v(t)\, ds\, dt$$

$$= 2 \int_a^T M(k)v(t)\, dt,$$

(13.11.2)

where $M(k)$ $M(k, t)$.

13.12.   Functional Equation for $q(t, s, a)$

Using this expression in (13.10.6), we obtain the relation

$$\frac{\partial f}{\partial a} = \left(v(a) + \int_a^T M(k)v(t)\, dt/2 \right)^2$$

$$= v(a)^2 + v(a) \int_a^T M(k)v(t)\, dt + \int_a^T \int_a^T \frac{M(k, t_1)M(k, t_2)v(t_1)v(t_2)\, dt_1\, dt_2}{4}$$

(13.12.1)

On the other hand, using (13.9.2),

$$\frac{\partial f}{\partial a} = v(a)^2 - 2v(a) \int_a^T q(a, s, a)v(s)\, ds + \int_a^T \int_a^T \frac{\partial q}{\partial a}(t, s, a)v(t)v(s)\, dt\, ds.$$

(13.12.2)

Comparing (13.12.1) and (13.12.2), which holds for all $v \in L^2(a, T)$, we obtain the two relations

$$q(a, t, a) = k(a, t) - \int_a^T q(t, w, a)k(a, w)\, dw$$

(13.12.3)

and

$$\frac{\partial q}{\partial a}(t, s, a) = \frac{M(k, t)M(k, s)}{4}$$

$$= q(a, t, a)q(a, s, a).$$

(13.12.4)

The last relation is the desired analog of the Riccati equation in the finite-dimensional case of ordinary differential equations.

EXERCISE

1.   Can (13.12.4) be used computationally, starting with the fact that it is easy to solve the Fredholm integral equation for $T - a \ll 1$?

13.13. The Potential Equation

Consider the potential equation

$$u_{xx} + u_{yy} = 0, \qquad (x, y) \in R,$$
$$u(x, y) = g(x, y), \qquad (x, y) \in B,$$
(13.13.1)

where B is the boundary of R. The partial differential equation is the Euler equation of the Dirichlet functional

$$J(u) = \int_R (u_x^2 + u_y^2) \, dA.$$
(13.13.2)

It is tempting then to attempt a computational solution of (13.12.1) applying the dynamic programming approach to the minimization of $J(u)$. Starting with R, we cut slices off it successively (see Figure 13.1), regarding $g(p)$, the values of $g(x, y)$ along one side of R, as the state vector.

This procedure works very well in practice. Detailed discussions will be found in papers cited at the end of this chapter.

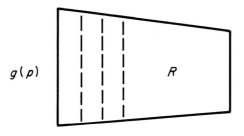

Figure 13.1

13.14. Reduction of Dimensionality—I

As we have mentioned above, the major hurdle to a computational treatment of distributed control processes resides in the dimensionality of the associated state vectors. In some cases we can use various devices to reduce this dimensionality. One is the method already discussed in Chapter 11.

Suppose that the problem is that of minimizing the functional

$$J(u) = \sum_{k=1}^{K} w_i g(u(x_i, T)) + \int_0^T \int_0^1 h(v) \, dx \, dt,$$
(13.14.1)

where

$$u_t = u_{xx} + v, \qquad t > 0, \qquad 0 < x < 1,$$

$$u(x, 0) = c(x), \qquad\qquad 0 < x < 1,$$

$$u(0, t) = u(1, t) = 0, \qquad\qquad t > 0. \qquad (13.14.2)$$

The fact that we can write

$$u(x, t) = w(x, t) + \int_0^1 dy \int_0^t k(x, y, t - s)v(y, s) \, ds \qquad (13.14.3)$$

enables us to convert the problem via dynamic programming into a K-dimensional one.

A similar reduction can be performed in the case where it is desired to minimize the functional

$$J(u) = g(u(T), u'(T), \ldots, u^{(k-1)}(T)) + \int_1^T h(v) \, dt, \qquad (13.14.4)$$

where

$$u'(t) = a_1 u(t) + a_2 u(t - 1) + v(t), \qquad t \geq 0,$$

$$u(t) = c(t), \qquad\qquad 0 \leq t \leq 1. \qquad (13.14.5)$$

EXERCISES

1. Why is the reduction of dimensionality important even in the "easy" cases were g and h are quadratic?
2. Obtain the dynamic programming relations in these cases.

13.15. Reduction in Dimensionality—II

In general, we cannot expect to find an analytic transformation which provides an exact reduction of dimensionality. Instead, as usual, some methods of approximation must be used. Sometimes these will be in state space, sometimes in policy space, other times in function space and structure space. References to some work of this nature will be found at the end of this chapter.

Miscellaneous Exercises

1. Show that the results of (13.12.4) can be obtained in the following fashion without the aid of variational techniques.
   (a) Starting with $u(t) = v(t) + \int_a^1 k(t, s)u(s) \, ds$, we have

   $$u(t) = v(t) + \int_a^1 K(t, s\, a)v(s) \, ds.$$

(b)   Hence, $u_a(t) = K(t, a, a)v(a) + \int_a^1 \dfrac{\partial K}{\partial a}(t, s, a)v(s)\, ds.$

(c)   The original equation yields $u_a(t) = -k(t, a)u(a) + \int_a^1 k(t, s)u_a(s)\, ds.$

(d)   Regarding this as an integral equation for $u_a(t)$, we have . . . . (See

R. Bellman, "On the Variation of the Fredholm Resolvent," *Boll. UMI*, **14**, 1959, pp. 78–81.)

2.   Consider the equation $u'' + (p(x) + tq(x))u = v(x)$, $u(0) = u(1) = 0$, with the solution $u(x) = \int_0^1 K(x, y, t)(y)\, dy$. Show that

$$\frac{\partial K}{\partial t}(x, y, t) = -\int_0^1 K(x, z, t)K(z, y, t)q(z)\, dz.$$

(M. Schiffer)

BIBLIOGRAPHY AND COMMENT

13.1.   See

R. Bellman and R. Kalaba, "Dynamic Programming Applied to Control Processes Governed by General Functional Equations," *Proc. Nat. Acad. Sci. USA*, **48**, 1962, pp. 1735–1737.

R. Bellman and H. Osborn, "Dynamic Programming and the Variation of Green's Functions," *J. Math. Mech.*, 7, 1958, pp. 81–86.

D. Mangeron, "Sur une classe d'equations fonctionnelles attaches aux problems extremaux correspondant aux systemes polyvibrants," *C. R. Acad. Sci. Paris*, **266**, 1968, pp. 1121–1124.

D. Mangeron, "The Bellman Equations of Dynamic Programming Concerning a New Class of Boundary Value Problems with Total Derivatives," *J. Math. Anal. Appl.*, 9, 1964, pp. 141–146.

P. K. C. Wang, "Optimal Control of Distributed Control Systems with Time Delays," *IEEE Trans. Automat. Control*, **AC-9**, 1964, pp. 13–22.

P. K. C. Wang, "Control of Distributed Parameter Systems," in *Advances in Control Systems*, Volume 1, Academic Press, New York, 1964, pp. 75–172.

P. K. C. Wang and F. Tung, "Optimal Control of Distributed Parameter Systems," J. Basic Engrg., *TRANS. ASME*, Ser. D, **86**, 1964, pp. 67–79.

D. M. Wiberg, "Feedback Control of Linear Distributed Systems," *Trans. ASME*, **89**, 1967, pp. 379–384.

13.7.   See the papers cited in Section 13.1.

13.8.   This was given in

R. Bellman, "Functional Equations in the Theory of Dynamic Programming—VII: A Partial Differential Equation for the Fredholm Resolvent," *Proc. Amer. Math. Soc.*, **8**, 1957, pp. 435–440.

13.13.   These ideas are expounded in detail in a series of papers,

E. Angel, "Dynamic Programming and Partial Differential Equations," USCEE-265, Univ. of Southern California, Los Angeles, March 1968.

E. Angel, "A Building Block Technique for Elliptic Boundary-value Problems over Irregular Regions," USCEE-285, Univ. of Southern California, Los Angeles, June 1968.

E. Angel, "Discrete Invariant Imbedding and Elliptic Boundary-value Problems over Irregular Regions," USCEE-279, Univ. of Southern California, Los Angeles, June 1968.

E. Angel, "Noniterative Solutions of Nonlinear Elliptic Equations," IUSCEE-306, Univ. of Southern California, Los Angeles, October 1968.

E. Angel, "Invariant Imbedding and Three-dimensional Potential Problems," USCEE-325, Univ. of Southern California, Los Angeles, January, 1969.

E. Angel, "Inverse Boundary-value Problems: Elliptic Equations," USCEE-343, Univ. of Southern California, Los Angeles, April 1969.

13.14. See

R. Bellman, "Terminal Control, Time Lags, and Dynamic Programming," *Proc. Nat. Acad. Sci. USA*, **43**, 1957, pp. 927–930.

R. Bellman and R. Kalaba, "Reduction of Dimensionality, Dynamic Programming, and Control Processes," *J. Basic. Eng.*, March 1961, pp. 82–84.

13.15. See

R. Bellman, "A New Type of Approximation Leading to Reduction of Dimensionality in Control Processes," *J. Math. Anal. Appl.*, **27**, 1969, pp. 454–459.

A. Lew, "Reduction of Dimensionality by Approximation Techniques: Diffusion Processes," *J. Math. Anal. Appl.*, to appear.

14

SOME DIRECTIONS OF RESEARCH

14.1. Introduction

The reader who has resolutely followed our path through the foregoing pages will probably require little convincing that a great deal remains to be done in those areas of control theory which we have examined in so cursory a fashion. Nonetheless, there may remain some subliminal fears that all of the really interesting processes have been already formulated. Perseverence and hard work are required, to be sure, to draw the completed map, but perhaps little remains for the intrepid explorer.

This is fortunately not at all the case. If anything, precisely the opposite holds. The greater part of control theory is terra incognita. To demonstrate this basic point, one of the major appeals of this field, we shall briefly indicate a number of classes of processes which have either not been mentioned at all or, if so, then touched on very lightly. A number of these are the subjects of current research papers; they will be treated in Volumes III and IV. Many require considerable effort in formulation.

We have repeatedly played on the theme of computational feasibility and emphasized its importance as a guide to research. Constant attention to these matters generates new mathematical problems and often results in the construction of analytic approaches, which are entirely different from the original. Closely connected with this purely mathematical method for the discovery of significant areas of investigation is the idea that questions of engineering and physical feasibility are equally stimulating. In particular, we want to keep in mind that the control of a real system requires both devices and people. We need devices that can sense, communicate, process data, make decisions and implement decisions; we need people who can observe, diagnose, interpret, countermand decisions, choose criteria, and teach. Very little has been done in the crucial field of man–machine control systems, where the interplay of different duties and skills is crucial.

Any careful analysis of the roles and interactions of the different components of an operating system enables us to formulate many novel and interesting classes of mathematical processes. We accept as an axiom that a meaningful system produces meaningful mathematics. This is not the only way to create good mathematics, but it is an easy way, which is guaranteed to succeed.

14.2. Constraints

In the previous chapters we have formulated a continuous control process in the following form: Determine the minimum over y of the functional

$$J(x, y) = \int_0^T h(x, y) \, dt, \tag{14.2.1}$$

where x and y are connected by the differential equation

$$x' = g(x, y), \qquad x(0) = c. \tag{14.2.2}$$

We indicated that global constraints

$$\int_0^T k(x, y) \, dt \le r_1, \tag{14.2.3}$$

could sometimes be expeditiously handled by the use of a Lagrange multiplier without seriously affecting methodologies based on the use of the Euler equation. However, introduction of local restrictions on the state and control vector, e.g.,

(a) $|x(t)| \le a_1,$ $0 \le t_0 < t \le T_0 \le T$, or

(b) $|y(t)| \le a_2,$ over a similar region, or

(c) $x' - g_1(x, y) \le a_3(t),$ (14.2.4)

leads to interesting new kinds of analytic and computational questions in both the calculus of variations and dynamic programming. In general, the set of allowable states and decisions will depend on the history of the systems and, in particular, upon the previous decisions. For example, t_0 and T_0 in (14.2.4) could depend on t and x.

Variational processes of this nature were studied many years ago, but they received little sustained attention until the exigencies of economic and engineering decision-making forced a renewed interest in obtaining analytic and computational results and until computers were available to treat meaningful problems.

In using the criterion function (14.2.1) we make the blithe assumption that the duration of the process, T, is fixed, independent of the control vector or policy. This is a sensible initial assumption which in the main is not valid for any control process of significance. We expect the duration of the process to depend on both the nature of the control exerted and the behavior of the system. We encounter, in this fashion, a functional of the form

$$J(x, y) = \int_0^{T(x, y)} h(x, y) \, dt \qquad (14.2.5)$$

As an example, let us cite the problem of "soft landing" on the moon—the minimization of the speed at the time of impact or the problem of "minimum miss distance." In both cases, the time of evaluation, T, depends on the policy employed.

Oddly, "stop rules" have been the subject of intensive investigation in the area of stochastic control processes, but not in the field of deterministic control processes, for some time. Questions of this nature are of "free boundary" type, problems hitherto investigated principally in the physical domain, e.g., hydrodynamics, aerodynamics, the theory of melting, and so on.

14.3. Control Takes Time

A first consequence of the fact that systems themselves are needed to control systems is that control takes time. Time is one of the most valuable resources we have; it is unique in the fact that it cannot be reversed or replaced. It takes time to sense, process and communicate, to make decisions and then to implement those decisions.

We are led by these considerations to introduce mathematical systems containing delays ("time-lags") and, more generally, hereditary effects. We are concerned with systems where the rate of change of state depends upon the past history of the system. One analytic formulation of control processes of this nature leads to differential-difference equations, or equations of the form

$$x'(t) = g(x(t), x(t - t_1), \ldots, x(t - t_k) : y(t), y(t - t_1), \ldots, y(t - t_k)) \quad (14.3.1)$$

As we have seen in a preceding chapter, the determination of optimal control encounters serious analytic and computational obstacles. These difficulties are compounded by the fact that a more critical examination of the control processes involved leads us first to replace (14.3.1) by

$$x'(t) = g\left(x(t), \quad \int_{-\infty}^{t} x(t-s)\, dh(s), \ldots \right), \qquad (14.3.2)$$

then by

$$x'(t) = g\left(x(t), \quad \int_{-\infty}^{t} x(t-s)\, dh(t, s), \ldots \right), \qquad (14.3.3)$$

(which is to say, the lags are not constant but time-dependent), and then by

$$x'(t) = g\left(x(t), \quad \int_{-\infty}^{t} x(t-s)\, dh(x, y, t, s), \ldots \right), \qquad (14.3.4)$$

(which is to say, the lags depend upon the time, the state of the system, and the nature of the control).

Many interesting questions arise in ascertaining the proper balance among the cost in time and other resources required for various operations, the accuracy attained, and, ultimately, the cost of decision. This in turn is part of a modern theory of experimentation, a theory not yet in existence.

14.4.   Principle of Macroscopic Uncertainty

All of this enables us to formulate a suitably vague general principle: *It is impossible to control a large system perfectly.*

The point is that a large system automatically requires an appreciable time for the essential steps of control. While these steps are being taken, the system is behaving in its own fashion. If we attempt to reduce the time for an individual step, we incur costs either in accuracy or in resources. Conversely, increased accuracy requires time and resources.

Clearly, there are many different mathematical questions associated with the foregoing principle, and it would be important to investigate them.

14.5.   "On-Line" Control

Closely associated with the foregoing problem area is that of "on-line" control. Here the constraint is a novel one mathematically—one not previously encountered in scientific research. We are required to render a decision, perhaps supply a numerical answer, within a specified period of time. It is no longer a question of devising a computationally feasible algorithm; instead, we must obtain the best approximation within a specified time. Questions of this nature arise within every operating system, and particularly in the medical field.

Many of the approximation methods discussed in the foregoing chapter can be employed. One idea which merits extensive investigation is that of "sufficient statistics." As the name indicates, the concept comes from the theory of stochastic processes. The basic idea is to reduce dimensionality, and thus time in the determination of optimal policies, by considering only those combinations of state variables and parameters which suffice for the determination of the optimal policy within a specified accuracy. Little has been done in applying this fundamental notion to deterministic processes.

As we know from mathematical physics, the asymptotic behavior of a system may be quite simple and reasonable despite the complexity of transient or short-term, behavior. It follows that in general there will exist asymptotic sufficient statistics which can be used for long-term control. This is part of the general theory of asymptotic control, which was barely touched upon in an earlier chapter.

14.6. Monotone Approximation

Sometimes, we are fortunate in being able to design a system from the start. In general, the applications of control theory are to on-going systems which can only be modified in certain ways. Little attention has been paid to how one introduces improved techniques into an operating system, subject to these constraints, and, in particular, how these techniques can be introduced so as to improve performance in a steady fashion with no undesirable, temporary deterioration of the system.

In general, there has been little study of successive approximation guaranteeing monotonicity.

14.7. Identification and Control

In many cases we are given a descriptive equation or the form

$$x' = g(x, a), \qquad x(0) = c, \tag{14.7.1}$$

where the structure of the function g is known, but the parameters a and c may be unknown. For example, we may know that the equation is linear, $x' = Ax$, $x(0) = c$, but the matrix A may not be given. A basic problem of identification is that of determining a and c on the basis of observations, a set of values $\{(x(t_i), b_i)\}$, $i = 1, 2, \ldots, M$. The theory of quasilinearization has been extensively used for this purpose.

Let us consider an extension of this problem. Suppose that

$$x' = g(x, y, a), \qquad x(0) = c,$$

$$J(x, y) = \int_0^T h(x, y) \, dt \tag{14.7.2}$$

and that it is desired to minimize J subject to the condition that $(x(t_1), b_1)$ is made available at time t_1, $(x(t_2), b_2)$ at time t_2, and so on. On-line identification processes are important in the biomedical field, but they have not been studied to any extent.

14.8. Mathematical Model-Making

The foregoing is part of a more general study of mathematical model-making as a control process. A major problem is that of balancing physical and mathematical complexity. Often in the consideration of simple versions of physical processes, we encounter major analytical obstacles; frequently important physical processes can be effectively dealt with by using low level mathematical processes, particularly if a digital computer is available. We don't understand the relations between different types of complexity.

Related to this is the task of devising useful notations for self-communication, communication with others and with a digital computer.

14.9. Computing as a Control Process

We have already referred to the fact that computing can be considered to be a control process in which we want to blend complexity, time, and accuracy in some appropriate way. It should, in addition, be treated in many cases as an adaptive control process in which the results of the previous calculation are used to guide the subsequent calculations, producing not only a choice of parameters but a choice of algorithms.

14.10. Physiological Control Processes

Over the past 25 years the major influences upon the development of control theory have been the demands of engineering, with economic and industrial processes occupying a secondary role. It is safe to predict that the situation will change drastically over the next 5 years. The biomedical fields will become the paramount sponsors of mathematical research in control theory and numerous other fields as well.

Let us, for example, mention the problem of an artificial heart, the determination of optimal drug administration (part of the burgeoning field of pharmacokinetics), and the many problems arising in sensory prosthesis.

14.11. Environmental Control

The study of large scale systems will play a major role in the control of air and water pollution, in the "farming of the oceans" and in the never-ending battle against the insect world. As the population of the world increases, this battle for survival *la lutte pour la vie*, will intensify.

In studies of this nature, we meet the challenge of incommensurability, vector-valued criteria. We can expect the new theory of "fuzzy systems" to play a vital role in this area.

14.12. Classical and Nonclassical Mechanics

Almost all of existing control theory—deterministic, stochastic and adaptive—is cast in the framework of classical mechanics. We need a relativistic and quantum mechanical control theory to handle many of the new processes of technology , and we need a statistical mechanics for control theory. In any case, the mathematical challenges are fascinating.

14.13. Two-Person and *N*-Person Processes

In economic and military processes we encounter control processes where the decision-makers are partially or totally in opposition. We have not discussed processes of this nature here because the most interesting versions require the theory of probability and a careful analysis of the concept of information.

14.14. Conclusion

Our aim has been to briefly indicate the unlimited opportunities for research within the domain of deterministic control processes. As soon as we allow stochastic and adaptive features, it becomes a major task even to catalogue a portion of the problem area. It is safe to conclude that control theory will be a field of major significance for as far as we can see in the clouded crystal ball.

BIBLIOGRAPHY AND COMMENTS

14.1.   General background reading, with many additional references, is

R. Bellman, Adaptive Control Processes ...
R. Bellman, Some Vistas of Modern Mathematics, ...
R. Bellman, *Dynamic Programming: A Reluctant Theory*, *New Methods of Thought and Procedure*, Springer, Berlin, pp. 99–123.
R. Bellman, and P. Brock, "On the Concepts of a Problem and Problem-Solving," *Amer. Math. Monthly*, 67, 1960, pp. 119–134.

14.2.   For a discussion of variatonal problems subject to constraints, see

R. Bellman, "Dynamic Programming and Its Application to the Variational Problems in Mathematical Economics," *Proc. Symp. Calculus of Variations and Appli.*, April 1956, pp. 115–138.
H. Hermes and J. P. La Salle, *Functional Analysis and Time Optimal Control*, Academic Press, New York, 1969.

and the books by

R. Bellman, I. Glicksberg and O. Gross, M. Hestenes and L. Pontrjagin previously cited.

The detailed study of the variational problems subject to constraints given in the Bellman, Glicksberg and Gross work leads to the use of dynamic programming to obtain computationally effective algorithms. Prior to this, the theory had been used exclusively for stochastic processes.

14.3.   See the book by R. Bellman and K. L. Cooke previously cited and

R. Bellman, J. Buell and R. Kalaba, "Mathematical Experimentation in Time-lag Modulation," *Comm. Assoc. Comput. Machinery*, 9, 1966, p. 752.

14.4.   See

A Survey of the Status and Trends in Large Scale Control Systems Theory and Applications, Office of Central Theory and Applications, NASA, March, 1970.

14.8.   See

R. Bellman, "Mathematical Model-making as an Adaptive Control Process," in *Mathematical Optimization Techniques*, Univ. of California, Press, 1963, pp. 333–339.

14.9.   See

D. E. Knuth, *The Art of Computer Programming*, Volume I: *Fundamental Algorithms*, Addison-Wesley, 1968.

14.10.   See, for example,

R. Bellman, ed., *Proc. Symp. Appl. Math.* in *Mathematical Problems in the Biological Sciences*, Vol. 14. American Mathematical Society, 1962.
P. I. Gulyayev, "The Significance of Dynamic Programming and Theory of Games for Physiology," (in Russian with English summary), *Colloq. Nervous System*, 3, LGU, Leningrad, 1962, pp. 177–189.
R. E. Kalman, "New Developments in Systems Theory Relevant to Biology, Systems Theory and Biology," *Proc. III Sys. Symp. at Case Institute of Technology*, Springer, Berlin, 1968.

M. A. Khvedelidze, "Investigation of the Control Processes in Living Organisms and Ways of Developing New Cybernetic Systems," (in Russian), *Tr. Inst. Kibernet., AN Gruz SSSR*, **1**, 1963, pp. 169–190

R. B. Laning, *Studies Basic to the Consideration of Artificial Heart Research and Development Program*, Final Rep. PB 169–831, U.S. Dept. of Commerce, National Bureau of Standards, Inst. for Applied Technology, Washington, D.C., 1961.

L. D. Liozner, "Regeneration of Lost Organs," *Izdatel's stvu Akad. Nauk SSSR*, Moscow 1962, pp. 1–141; English transl., FTD-TT 63–576, Div. Foreign Technology, Air Force Systems Command, Wright-Patterson Air Force Base, Sept. 1963.

N. Wiener, "Problems of Sensory Prosthesis," *Bull. Amer. Math. Soc.*, **57**, no. 1—part 1, Jan. 1951.

R. Bellman, J. Jacquez and R. Kalaba, "The Distribution of a Drug in the Body," *Bull. Math. Biophys.*, **22**, 1960, pp. 309–322.

R. Bellman, J. Jacquez and R. Kalaba, "Some Mathematical Aspects of Chemotherapy—I: One-Organ Models," *Bull. Math. Biophys.*, **22**, 1960, pp. 181–198.

E. I. Jury and T. Pavlidis, "A Literature Survey of Biocontrol Systems," *IEEE Trans. on Autom. Control*, **AC-8**, no. 3, July 1963, pp. 210–217.

E. Kruger-Thiemer and R. R. Levine, "The Solution of Pharmacological Problems with Computers," *Drug. Res.*, **18**, 1968, pp. 1575–1579.

R. Tomovic, and G. Boni, "An Adaptive Artificial Hand," *IRE Trans. on Autom. Control*, **AC-7**, no. 3, April 1962, pp. 3–10.

R. Tomovic and L. Radanopvic, "Homeostatic Control of Dynamic Systems, "*Proc. First Int. Symp. Optimizing Adap. Control* (L. E. Bollinger, E. J. Minnar, and J. G. Truxal, eds.), sponsored by *Theory Comm. IFAC, April* 1962, pp. 57–67.

14.11. See

P. A. Larkin, "The Possible Shape of Things to Come," *SIAM Rev.*, **11**, 1969, pp. 1–6.

L. B. Slobodkin, "The Strategy of Evolution," *Amer. Sci.*, **52**, 1964, pp. 342–357.

K. E. F. Watt, "Dynamic Programming, Look Ahead Programming and the Strategy of Insect Pest Control," *Can. Entomol.*, **95**, 1963, pp. 525–536.

K. E. F. Watt, "Computers and the Evaluation of Resource Management Strategies," *Amer. Sci.*, **52**, 1964, pp. 408–418.

14.12. See

H. G. L. Krause, *Astrorelativity*, NASA Tech. Rep. TR R-188, January 1964.

14.13. See

F. Bellman, *Dynamic Programming*, Princeton Univ. Press, Princeton, N.J., 1957.

L. D. Berkovitz, *A Variational Approach to Differential Games*, Study no. 52, in *Annals of Mathematics*, Princeton University Press, Princeton, N.J., 1964.

AUTHOR INDEX

A

Aczel, J., 39
Altshuler, S. 137
Angel, E., 67, 288
Aoki, M., 68, 135, 222, 258
Aris, R., 37, 66, 246
Arnold, R. F., 37
Aronsson, G., 107
Arrow, K. J., 106
Ash, M., 39
Athans, M., 137
Avriel, M., 69
Azen, S. P., 68, 165

B

Babushka, I., 67
Bagwell, J. R., 34
Bailey, P. B., 135
Baldwin, J. F., 165
Banks, H. T., 276
Beckenbach, E. F., 39, 106, 165, 224
Beckwith, R. E., 166, 180

Beightler, C. S., 34, 66, 67
Bellman, R., 8, 9, 33, 34, 35, 36, 37, 38, 39, 66, 67, 68, 69, 104, 105, 106, 107, 135, 136, 137, 161, 162, 163, 164, 165, 166, 180, 201, 222, 223, 224, 225, 237, 245, 246, 257, 258, 276, 277, 287, 288, 296, 297
Berkovich, E. M., 180
Berkowitz, L. D., 163, 297
Bertele, U., 37, 66
Boltianskii, V. G., 163, 164
Boni, G., 297
Borch, K., 9
Borel, Emil, 264
Brioschi, F., 37, 66
Brock, P., 9, 296
Brown, T. A., 37, 39
Bryan, G. L., 9
Bryson, A. E., 137
Bucy, R., 201
Budak, B. M., 180
Buell, J., 296
Burkov, V. N., 246
Butkowski, A. G., 276
Butler, T., 39
Butz, A. R., 66

SUBJECT INDEX

Mathematics in Science and Engineering

A Series of Monographs and Textbooks

Edited by RICHARD BELLMAN, *University of Southern California*

39. Y. Sawaragi, Y. Sunahara, and T. Nakamizo. Statistical Decision Theory in Adaptive Control Systems. 1967

40. R. Bellman. Introduction to the Mathematical Theory of Control Processes Volume I. 1967; Volume II. 1971 (Volume III in preparation)

41. E. S. Lee. Quasilinearization and Invariant Imbedding. 1968

42. W. Ames. Nonlinear Ordinary Differential Equations in Transport Processes. 1968

43. W. Miller, Jr. Lie Theory and Special Functions. 1968

44. P. B. Bailey, L. F. Shampine, and P. E. Waltman. Nonlinear Two Point Boundary Value Problems. 1968

45. Iu. P. Petrov. Variational Methods in Optimum Control Theory. 1968

46. O. A. Ladyzhenskaya and N. N. Ural'tseva. Linear and Quasilinear Elliptic Equations. 1968

47. A. Kaufmann and R. Faure. Introduction to Operations Research. 1968

48. C. A. Swanson. Comparison and Oscillation Theory of Linear Differential Equations. 1968

49. R. Hermann. Differential Geometry and the Calculus of Variations. 1968

50. N. K. Jaiswal. Priority Queues. 1968

51. H. Nikaido. Convex Structures and Economic Theory. 1968

52. K. S. Fu. Sequential Methods in Pattern Recognition and Machine Learning. 1968

53. Y. L. Luke. The Special Functions and Their Approximations (In Two Volumes). 1969

54. R. P. Gilbert. Function Theoretic Methods in Partial Differential Equations. 1969

55. V. Lakshmikantham and S. Leela. Differential and Integral Inequalities (In Two Volumes). 1969

56. S. H. Hermes and J. P. LaSalle. Functional Analysis and Time Optimal Control. 1969

57. M. Iri. Network Flow, Transportation, and Scheduling: Theory and Algorithms. 1969

58. A. Blaquiere, F. Gerard, and G. Leitmann. Quantitative and Qualitative Games. 1969

59. P. L. Falb and J. L. de Jong. Successive Approximation Methods in Control and Oscillation Theory. 1969

60. G. Rosen. Formulations of Classical and Quantum Dynamical Theory. 1969

61. R. Bellman. Methods of Nonlinear Analysis, Volume I. 1970

62. R. Bellman, K. L. Cooke, and J. A. Lockett. Algorithms, Graphs, and Computers. 1970

63. E. J. Beltrami. An Algorithmic Approach to Nonlinear Analysis and Optimization. 1970

64. A. H. Jazwinski. Stochastic Processes and Filtering Theory. 1970

65. P. Dyer and S. R. McReynolds. The Computation and Theory of Optimal Control. 1970

66. J. M. Mendel and K. S. Fu (eds.). Adaptive, Learning, and Pattern Recognition Systems: Theory and Applications. 1970

67. C. Derman. Finite State Markovian Decision Processes. 1970

68. M. Mesarovic, D. Macko, and Y. Takahara. Theory of Hierarchial Multilevel Systems. 1970

69. H. H. Happ. Diakoptics and Networks. 1971

70. Karl Astrom. Introduction to Stochastic Control Theory. 1970

71. G. A. Baker, Jr. and J. L. Gammel (eds.). The Padé Approximant in Theoretical Physics. 1970

72. C. Berge. Principles of Combinatorics. 1971

73. Ya. Z. Tsypkin. Adaptation and Learning in Automatic Systems. 1971

74. Leon Lapidus and John H. Seinfeld. Numerical Solution of Ordinary Differential Equations. 1971

75. L. Mirsky. Transversal Theory, 1971

In preparation

Harold Greenberg. Integer Programming

E. Polak. Computational Methods in Optimization: A Unified Approach

Thomas G. Windeknecht. A Mathematical Introduction to General Dynamical Processes

Andrew P. Sage and James L. Melsa. System Identification

R. Boudarel, J. Delmas, and P. Guichet. Dynamic Programming and Its Application to Optimal Control

William Stenger and Alexander Weinstein. Methods of Intermediate Problems for Eigenvalues Theory and Ramifications